INDOLE ALKALOIDS

Visual Guides to Natural Product Synthesis

INDOLE ALKALOIDS
SPIROOXINDOLE

MALIHA UROOS
Centre for Research in Ionic Liquids, School of Chemistry, University of the Punjab, Lahore, Pakistan

ABDUL HAMEED
Department of Chemistry, University of Sahiwal, Sahiwal, Pakistan

SADIA NAZ
Centre for Research in Ionic Liquids, School of Chemistry, University of the Punjab, Lahore, Pakistan

MUHAMMAD RAZA SHAH
H. E. J. Research Institute of Chemistry, International Center for Chemical and Biological Sciences, University of Karachi, Karachi, Pakistan; Department of Chemistry, Forman Christian College (A Chartered University), Lahore, Pakistan

ELSEVIER

Elsevier
Radarweg 29, PO Box 211, 1000 AE Amsterdam, Netherlands
The Boulevard, Langford Lane, Kidlington, Oxford OX5 1GB, United Kingdom
50 Hampshire Street, 5th Floor, Cambridge, MA 02139, United States

Copyright © 2022 Elsevier Inc. All rights reserved.

No part of this publication may be reproduced or transmitted in any form or by any means, electronic or mechanical, including photocopying, recording, or any information storage and retrieval system, without permission in writing from the publisher. Details on how to seek permission, further information about the Publisher's permissions policies and our arrangements with organizations such as the Copyright Clearance Center and the Copyright Licensing Agency, can be found at our website: www.elsevier.com/permissions.

This book and the individual contributions contained in it are protected under copyright by the Publisher (other than as may be noted herein).

Notices
Knowledge and best practice in this field are constantly changing. As new research and experience broaden our understanding, changes in research methods, professional practices, or medical treatment may become necessary.

Practitioners and researchers must always rely on their own experience and knowledge in evaluating and using any information, methods, compounds, or experiments described herein. In using such information or methods they should be mindful of their own safety and the safety of others, including parties for whom they have a professional responsibility.

To the fullest extent of the law, neither the Publisher nor the authors, contributors, or editors, assume any liability for any injury and/or damage to persons or property as a matter of products liability, negligence or otherwise, or from any use or operation of any methods, products, instructions, or ideas contained in the material herein.

British Library Cataloguing-in-Publication Data
A catalogue record for this book is available from the British Library

Library of Congress Cataloging-in-Publication Data
A catalog record for this book is available from the Library of Congress

ISBN: 978-0-323-91674-5

For Information on all Elsevier publications
visit our website at https://www.elsevier.com/books-and-journals

Publisher: Susan Dennis
Editorial Project Manager: Mica Ella Ortega
Production Project Manager: R. Vijay Bharath
Cover Designer: Christian J. Bilbow

Typeset by MPS Limited, Chennai, India

Contents

List of figures	vii
List of schemes	ix
About the authors	xiii
Preface	xv

1. Introduction to spirooxindoles   1

References	4

2. (−)-Affinisine   5

2.1 Fonseca's first stereospecific total synthesis of (−)-affinisine (2014)	6
2.2 Improved total synthesis of (−)-affinisine (2017)	8
References	11

3. (+)-Austamide   13

3.1 Hutchison's total synthesis of (±)-austamide (1979)	14
3.2 Corey's total synthesis of (+)-austamide (2000)	18
References	20

4. Brevianamide A and B   23

4.1 Total syntheses of brevianamides A and B	25
References	33

5. Citrinadin A and B   35

5.1 Syntheses of citrinadins A and B spirooxindole core	37
5.2 Total syntheses of citrinadins A and B	43
References	55

6. Coerulescine and horsfiline   57

6.1 Total synthesis of coerulescine and horsfiline	58
References	84

7. Elacomine and isoelacomine   87

7.1 Total syntheses of (±)-elacomine and (±)-isoelacomine	88
References	97

vi Contents

8. Gelsemine 99

8.1 Total syntheses of gelsemine 100
8.2 Formal syntheses of gelsemine 122
References 134

9. Paraherquamide A and B 135

9.1 Total syntheses of paraherquamides A and B 136
9.2 McWhorter's formal synthesis of 6,7-dihydroxyoxindole; a subunit of paraherquamide A (1996) 145
9.3 Conversion of marcfortine A to paraherquamide A via paraherquamide B (1997) 146
References 149

10. Rhynchophylline and isorhynchophylline 151

10.1 Total syntheses of (−)-rhynchophylline and (+)-isorhynchophylline 152
10.2 Semisyntheses of (−)-rhynchophylline and (+)-isorhynchophylline 159
10.3 Formal syntheses of (−)-rhynchophylline and (+)-isorhynchophylline 161
References 171

11. Spirotryprostatin A 173

11.1 Total synthesis of spirotryprostatin A 174
References 182

12. Spirotryprostatin B 183

12.1 Total syntheses of (−)-spirotryprostatin B 184
References 199

13. Strychnofoline 201

13.1 Carreira's first total synthesis of (±)-strychnofoline (2002) 202
13.2 Carreira's racemic total synthesis of (±)-strychnofoline; highly convergent selective annulation reaction (2006) 204
13.3 Yu's enantioselective synthesis of (−)-strychnofoline (2018) 207
References 208

List of abbreviations 209
Index 213

List of figures

Figure 1.1	Structures of different spirooxindole natural alkaloids.	1
Figure 2.1	Structure of (−)-affinisine.	5
Figure 3.1	Structure of (+)-austamide.	13
Figure 3.2	Biosynthesis of austamide.	14
Figure 4.1	Structures of (+)-Brevianamides A and B.	23
Figure 4.2	Biosynthesis of brevianamides A and B.	24
Figure 5.1	Structures of citrinadins A and B.	36
Figure 5.2	Two plausible biogenetic precursors in the biosynthetic pathway of citrinadin core proposed by Kobayashi et al.	36
Figure 6.1	Structures of (−)-coerulescine and (−)-horsfiline.	57
Figure 7.1	Structures of (+)-elacomine and (−)-isoelacomine.	87
Figure 8.1	Structure of gelsemine.	99
Figure 9.1	Structures of paraherquamides A and B.	135
Figure 10.1	Structures of rhynchophylline and isorhynchophylline.	151
Figure 11.1	Structure of spirotryprostatin A.	173
Figure 12.1	Structure of (−)-spirotryprostatin B.	183
Figure 13.1	Structure of (−)-strychnofoline.	201

List of schemes

Scheme 2.1	Fonseca's first stereospecific total synthesis of (−)-affinisine	6
Scheme 2.2	Fonseca's improved total synthesis of (−)-affinisine	8
Scheme 3.1	Hutchison's total synthesis of (±)-austamide	15
Scheme 3.2	Corey's total synthesis of (+)-austamide	19
Scheme 4.1	Williams' asymmetric total synthesis of (−)-brevianamide B	26
Scheme 4.2	Lawrence's total synthesis of brevianamides A and B	28
Scheme 4.3	Scheerer's formal synthesis of brevianamide B	31
Scheme 5.1	Martin's formal synthesis of citrinadin A spirooxindole core	37
Scheme 5.2	Sorensen's synthesis of citrinadin B core architecture	39
Scheme 5.3	Wood's enantioselective synthesis of (+)-citranidin A core	41
Scheme 5.4	Martin's first enantioselective total synthesis of citrinadin A	44
Scheme 5.5	Synthesis of (−)-Citrinadin A isomer	48
Scheme 5.6	Synthesis of (+)-Citrinadin B	50
Scheme 5.7	Wood's enantioselective total synthesis of (+)-Citrinadin B	52
Scheme 6.1	Jones' total synthesis of horsfiline via radical cyclization	58
Scheme 6.2	Fuji's asymmetric total synthesis of (−)-horsfiline	61
Scheme 6.3	Carreira's racemic total synthesis of (±)-horsfiline	63
Scheme 6.4	Masanori's total synthesis of (±)-coerulescine	64
Scheme 6.5	Palmisano's asymmetric total synthesis of (−)-horsfiline	65
Scheme 6.6	Selvakumar's racemic total synthesis of (±)-coerulescine and (±)-horsfiline	66
Scheme 6.7	Murphy's racemic total synthesis of (±)-coerulescine and (±)-horsfiline	67
Scheme 6.8	Chang's racemic total synthesis of (±)-coerulescine	69
Scheme 6.9	Trost's asymmetric total synthesis of (−)-horsfiline	70
Scheme 6.10	Neuville and Zhu's total synthesis of horsfiline	72
Scheme 6.11	Maison's total synthesis of (±)-horsfiline	74
Scheme 6.12	Kulkarni's total synthesis of (±)-coerulescine and (±)-horsfiline	75
Scheme 6.13	White's racemic total synthesis of (±)-coerulescine and (±)-horsfiline	77

Scheme 6.14	Kim's total synthesis of (−)-coerulescine	79
Scheme 6.15	Comesse's racemic total synthesis of (±)-coerulescine	80
Scheme 6.16	Park's enantioselective total synthesis of (−)-horsfiline	81
Scheme 6.17	Hayashi's total synthesis of (−)-coerulescine and (−)-horsfiline	83
Scheme 6.18	Castillo's formal synthesis of (±)-coerulescine	84
Scheme 7.1	Borschberg's total synthesis of (±)-elacomine and (±)-isoelacomine	88
Scheme 7.2	Borschberg's total synthesis of (+)-elacomine and (−)-isoelacomine	89
Scheme 7.3	Horne's total synthesis of (±)-elacomine and (±)-isoelacomine	90
Scheme 7.4	White's total synthesis of (±)-elacomine	92
Scheme 7.5	Njardarson's asymmetric total synthesis of (+)-elacomine	94
Scheme 7.6	Takemoto's formal synthesis of (±)-elacomine and (±)-isoelacomine	95
Scheme 8.1	Fukuyama's first total synthesis of (±)-gelsemine	100
Scheme 8.2	Fukuyama's enantioselective total synthesis of (+)-gelsemine	106
Scheme 8.3	Overman's total synthesis of (±)-gelsemine	111
Scheme 8.4	Qin's total synthesis of (+)-gelsemine	115
Scheme 8.5	Qiu's total synthesis of (+)-gelsemine	119
Scheme 8.6	Johnson's synthesis of key tetracyclic gelsemine intermediate	123
Scheme 8.7	Johnson's spiroannelation of gelsemine	127
Scheme 8.8	Synthesis of spirocyclopentaneoxindole intermediate for (+)-gelsemine	128
Scheme 8.9	Zhou's second attempt for spirocyclopentaneoxindole intermediate through intramolecular Michael cyclization	132
Scheme 9.1	Williams' asymmetric stereocontrolled first total synthesis of paraherquamide A	136
Scheme 9.2	Williams and Cushing convergent, stereocontrolled, asymmetric total synthesis of paraherquamide B	142
Scheme 9.3	Synthesis of 6,7-dihydroxyoxindole; a subunit of paraherquamide A	146
Scheme 9.4	Conversion of marcfortine A to paraherquamide A via paraherquamide B	147
Scheme 10.1	Ban's first total synthesis of (±)-rhynchophylline and (±)-isorhynchophylline	152
Scheme 10.2	Hiemstra's total synthesis of (−)-rhynchophylline and (+)-isorhynchophylline	154
Scheme 10.3	Tong's formal synthesis of (−)-rhynchophylline and (+)-isorhynchophylline	156
Scheme 10.4	Tong's enantioselective total synthesis of isorhynchophylline	158

	List of schemes	xi

Scheme 10.5	Finch's semisynthesis of (−)-rhynchophylline and (+)-isorhynchophylline via oxidative transformations of indole alkaloid	159
Scheme 10.6	Brown's semisynthesis of (−)-rhynchophylline and (+)-isorhynchophylline starting from dihydrosecologanin aglycone	160
Scheme 10.7	Martin's racemic formal synthesis of (−)-rhynchophylline and (+)-isorhynchophylline via two ring-closing metathesis (RCM) reactions	161
Scheme 10.8	Itoh's enantioselective formal synthesis of isorhynchophylline via Mannich−Michael reaction	163
Scheme 10.9	Wang's enantioselective formal synthesis of (−)-rhynchophylline and (+)-isorhynchophylline	165
Scheme 10.10	Amat's enantioselective formal synthesis of *ent*-rhynchophylline and *ent*-isorhynchophylline	168
Scheme 10.11	Xia's racemic formal synthesis of (±)-rhynchophylline and (±)-isorhynchophylline	170
Scheme 11.1	Danishefsky's total synthesis of spirotryprostatin A	174
Scheme 11.2	Williams' asymmetric total synthesis of spirotryprostatin A	176
Scheme 11.3	Fukuyama's stereoselective total synthesis of spirotryprostatin A	178
Scheme 11.4	Gong's synthesis of diastereoisomers of spirotryprostatin A via asymmetric organocatalytic 1,3-dipolar cycloaddition	181
Scheme 12.1	Danishefsky's total synthesis of spirotryprostatin B	184
Scheme 12.2	Rosen's total synthesis of spirotryprostatin B via intramolecular Heck cyclization	186
Scheme 12.3	Biomimetic total synthesis of spirotryprostatin B by Ganesan	188
Scheme 12.4	Fuji's total synthesis of spirotryprostatin B via asymmetric nitroolefination	190
Scheme 12.5	Williams' asymmetric stereocontrolled total synthesis of spirotryprostatin B	192
Scheme 12.6	Horne's total synthesis of spirotryprostatin B	194
Scheme 12.7	Carreira's total synthesis of spirotryprostatin B3	195
Scheme 12.8	Trost's total synthesis of spirotryprostatin B via diastereoselective prenylation	198
Scheme 13.1	Carreira's first total synthesis of (±)-strychnofoline	202
Scheme 13.2	Carreira's racemic total synthesis of (±)-strychnofoline; highly convergent selective annulation reaction	204
Scheme 13.3	Yu's enantioselective synthesis of (−)-strychnofoline	207

About the authors

Dr. Maliha Uroos obtained her MSc (Chemistry) in 2003 from School of Chemistry, University of the Punjab. She completed her PhD in synthetic organic chemistry in 2012 from the University of Nottingham under HEC Faculty Development Scholarship. She was awarded EPSRC postdoctoral fellowship (2012–13) at the University of Nottingham, UK. She was appointed as a lecturer at the University of Punjab in 2007 and currently working as an associate professor (2020 to date) and Director of Centre for Research in Ionic Liquids, School of Chemistry, University of the Punjab. She has received the following awards: President Talent Forming Scholarship (2000), PhD Scholarship (2008), Leslie Crombie Prize UK for best PhD student (2011), and EPSRC Fellowship (2012). She has secured two major research grants from HEC under NRPU and TDF. She has several publications in leading international journals including patents and presented her work at various national and international conferences. Her main focus is synthetic organic chemistry, and research interests include synthesis and applications of ionic liquids, synthesis of biologically active molecules, sustainable energy production, green chemistry protocols, and nanochemistry.

Dr. Abdul Hameed is working in the area of organic synthesis. He has expertise in the area of synthetic chemistry. He did PhD in Organic Chemistry from the School of Chemistry, The University of Nottingham, UK. He worked as a Postdoc fellow at Max-Planck Institut für Polymerforschung, Germany. Recently, he received Gold Medal (2017/18) in the category for scientists under 40 in the field of chemistry by The Chemical Society of Pakistan. He has published many research papers in different peer-reviewed journals at national and international level. He is currently working in the field of organic synthesis, which includes the development of novel synthetic methodologies and their applications in the area of natural products synthesis, as well as the synthesis of various types of heterocyclic compounds to study their biological activities.

Ms. Sadia Naz obtained her BS Chemistry (4 years) from Lahore College for Women University, Lahore in 2016. After then, she joined School of Chemistry, University of the Punjab for MPhil studies (2016–18). Currently, she is a PhD student at Centre for Research in

Ionic Liquids, School of Chemistry, University of the Punjab (2018 to onwards) and is working as a research assistant in HEC NRPU project with Dr. Maliha Uroos (2019 to onwards). She has publications in international journals and has presented her work in various international and national conferences. Her research interests include synthetic organic chemistry, ionic liquids, green process developments for biorefineries, and conversion of agricultural waste materials into valuable platform chemicals.

Prof. Dr. Muhammad Raza Shah is working in the area of nanotechnology and supramolecular chemistry. He has authored three books (Elsevier) and also edited two books (Elsevier), along with seven chapters in books and more than 400 research articles in international journals with impact factor more than 1000. He has two US patents to his credit too. He has supervised and cosupervised 30 PhDs and 36 MS/MPhil scholars. The research projects of Dr. Shah is funded by both National and International funding agencies. He was declared top scientist for the year 2019 by the government of KPK. He received civil award *Tamgha-i-Imtiaz*-2015. Atta-ur-Rahman gold medal-2006 and Dr. Raziuddin gold medal-2015 by Pakistan Academy of Sciences. He received Abdus Salam (Nobel Laureate) award, and selected as TWAS Young Affiliate in 2010 by Third World Academy of Sciences. One of his authored books was declared the best book of the year 2017 by HEC. He received Avecena Science & Innovation Award 2015 by Brain Trust UK. He is a fellow Pakistan Academy of Sciences, fellow International Union of Pure and Applied Chemistry (IUPAC), and fellow Chemical Society of Pakistan. Prof. Shah remained a mentor of Pakistani International Chemistry Olympiad Team since 2008, and the team has won 21 bronze medals and 3 honorable mentioned. He is the Deputy Coordinator of National Chemistry Talent Contest Program of HEC. He is an editor of the *Journal of the Chemical Society of Pakistan* and Secretary General Chemical Society Pakistan. He served (2007–12) as a national representative in IUPAC. He was a member of National Nanotechnology Foresight Committee (PCST). He was a member of COMSTECH consultative working group of scientists from OIC countries, to prepare Ten-Year Plan of Action (TYPOA 2016–25). He is regularly organizing national and international workshops and delivering lectures. He conducted 17 Phase 1 clinical trials (BE-PK) along with 2 Phase II clinical trials for a multinational pharma as a PI. He led the vaccine (COVID-19) phase-1 clinical trial of Sinopharm company in Pakistan. He led the DRAP-approved clinical trial for the treatment of COVID-19 patients with traditional chinese medicines.

Preface

Natural product synthesis has always been a fascinating research area attributed to their unique structural architect and peculiar bioactivities. The journey of making these complex molecules via total synthesis is challenging and sometimes exhaustive, requiring innovative thinking and development of a new methodology. The comprehensive prior knowledge pertaining to collection of all synthetic efforts toward a particular core structure of interest could be of valuable assistance in this regard. This book is an addition to our visual guide series presenting the schematic total syntheses of natural products containing "spirooxindole" core structure.

Spirooxindole natural products are more than fascinating to synthetic chemists due to their vast structural diversity and broad biological spectrum. A number of spirooxindole motifs have been prepared both via stereoselective and racemic approach following multistep synthetic protocols. This book mounts up the synthetic efforts done, so far, for imperative spirooxindole molecules in perspicuous schemes highlighting the key steps involved in total, formal, and semisyntheses. This presentation mode will be helpful for synthetic chemists as well as pharmacists for a quick and easy overlook of previous synthetic strategies for a particular molecule and designing novel and advanced strategies. This spirooxindole visual guide volume will cover Brevianamide A and B, Citrinadin A and B, Coerulescine and Horsfiline, Elacomine and Isoelacomine, Gelsemine, Paraherquamide A and B, Rynchophylline, Isorynchophylline, Spirotryprostatin A and B, and Strychnofoline. These molecules have gained much attention of synthetic chemists, and a large number of synthetic approaches and total syntheses of these molecules have been reported so far. Thus the schematic compilation of natural product synthetic layouts in this book will provide an easy and speedy access to the readers to decode already developed synthetic routes toward the targeted spirooxindoles.

CHAPTER 1

Introduction to spirooxindoles

Oxindole alkaloids are a subclass of indole alkaloids, the first member of which was isolated from the roots of *Gelsemium sempervirens*, thus named as *Gelsemium* alkaloids. After that, many novel oxindole alkaloids were derived from *Mitragyna*, *Ourouparia*, *Rauwolfia*, *Aspidosperma*, and *Vinca*.[1]

Majority of these alkaloids belong to an interesting subclass of oxindoles; spirooxindoles. These are recognized by a common tryptamine-derived basic core and a unique spiro-pyrrolidine ring connection with the third position of oxindole framework (Fig. 1.1).

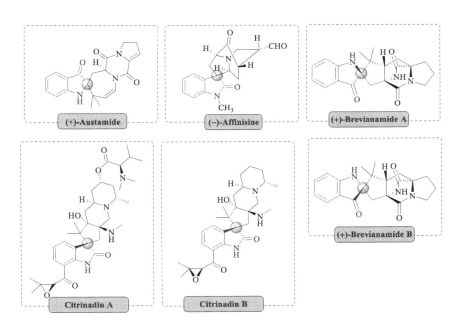

FIGURE 1.1 Structures of different spirooxindole natural alkaloids.

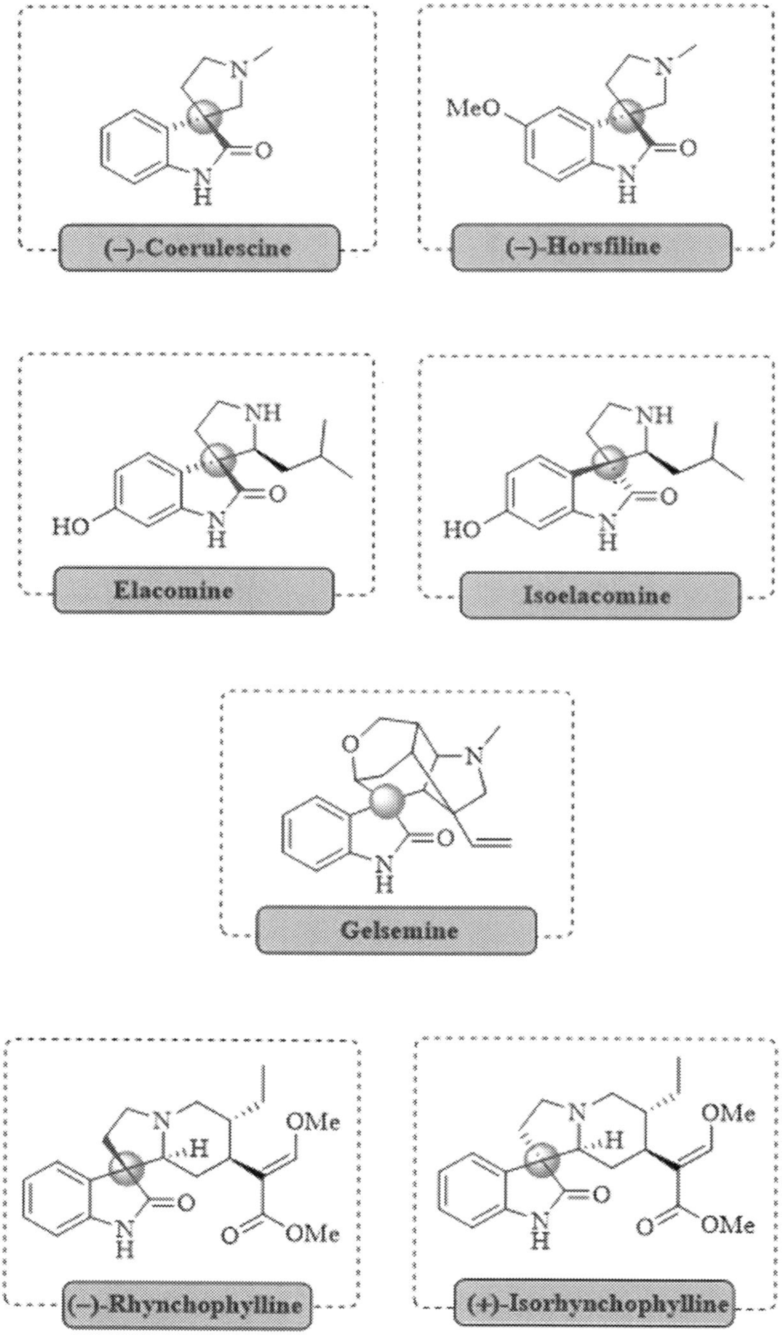

1. Introduction to spirooxindoles

FIGURE 1.1 (Continued).

Indole Alkaloids

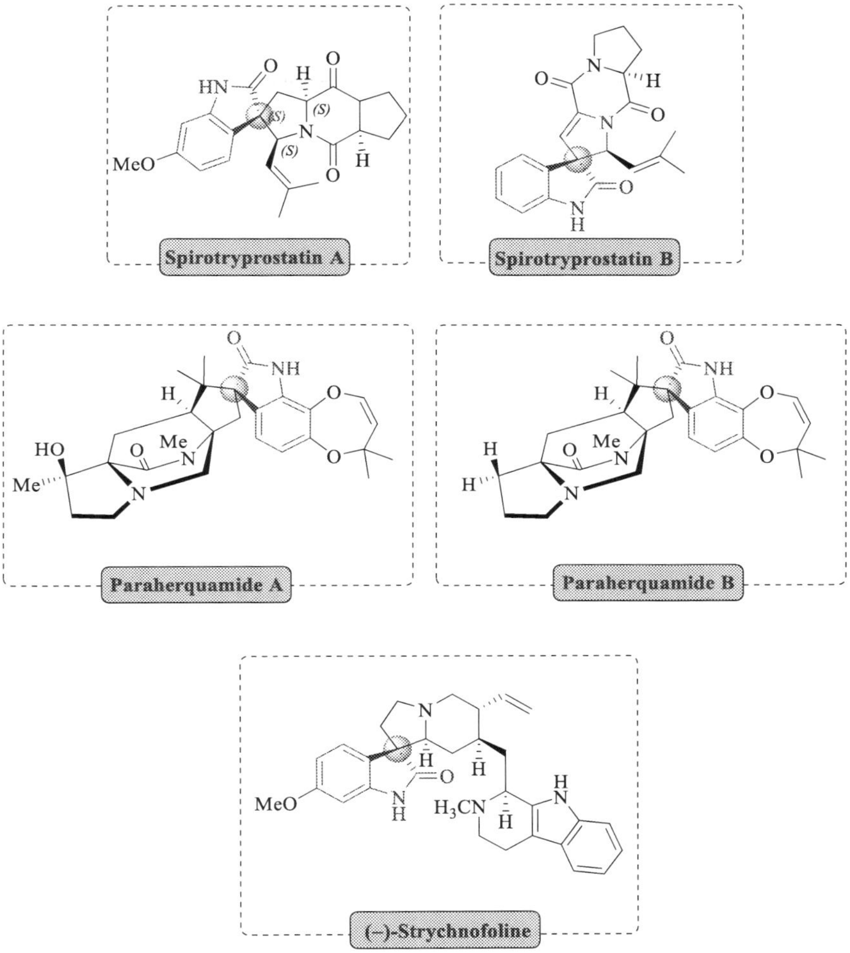

Figure contents:

Spirotryprostatin A

Spirotryprostatin B

Paraherquamide A

Paraherquamide B

(−)-Strychnofoline

FIGURE 1.1 (Continued).

Further classes of spirooxindoles based on their substructures involve tetracyclic secoyohimbane spirooxindoles such as rhynchophylline, pentacyclic heteroyohimbane spirooxindoles such as citrinadin, and some simplest prototype tricyclic artistic molecules with small substitutions on basic spirooxindole framework such as coerulescine, horsfiline, elacomine, and isoelacomine.

Spirooxindole alkaloids are of profound interest to both chemists and biologists as attractive targets for drug discovery due to their striking biological activities. For example, spirotryprostatin A is reported to

4

1. Introduction to spirooxindoles

inhibit the growth of G2M in mammalian tsFT210 cells,[2] strychnofoline inhibits various cell lines,[3] and horsfiline is an intoxicating snuff material.[4] Elacomine is reported to exhibit antitumor activity,[5] while rhynchophylline and spirotryprostatins A and B are recognized for their neuroprotective[6] and anticancer[2] behaviors, respectively. In fact, the spiro[oxindole-3,3'-pyrrolidine] is the most interesting framework constituting a pharmaceutically valuable class of biologically active compounds. Some other significant biological activities of these compounds include antimigraine activity[7] as well as contraceptive action.[8] In addition, nonplanar molecular geometry due to spiro-fusion also possesses a higher affinity to proteins 3D sites acting as bio-targets compared to the flat aromatic compounds.

In view of immense biological and pharmaceutical potential of these molecules, many efforts have been reported to develop their efficient synthetic routes. The most significant approaches involve multicomponent reactions for three-component condensation of amino acids, isatin and 1,3-dipolarophiles, Michael—Michael—aldol cascades, Heck reactions, and numerous other domino reactions.[9] Much attention has been paid to the regio- and stereoselectivities of these approaches to develop diversity-oriented synthetic protocols.

References

1. Bindra, J. In *The Alkaloids*; Manske, R. H. F., Ed.; Vol. 14; Academic Press: New York, 1973.
2. Cui, C.-B.; Kakeya, H.; Osada, H. Novel Mammalian Cell Cycle Inhibitors, Spirotryprostatins A and B, Produced by *Aspergillus fumigatus*, Which Inhibit Mammalian Cell Cycle at G2/M Phase. *Tetrahedron* **1996**, *52* (39), 12651−12666.
3. Dideberg, O.; Lamotte-Brasseur, J.; Dupont, L.; Campsteyn, H.; Vermeire, M.; Angenot, L. Structure Cristalline et Moléculaire d'un Nouvel Alcalöide Bisindolique: Complexe Moléculaire 1:2 Strychnofoline−Ethanol ($C_{30}H_{34}N_4O_2$. $2C_2H_6O$). *Acta Crystallogr., Sect. B: Struct. Sci. Cryst. Eng. Mater.* **1977**, *33* (6), 1796−1801.
4. Jossang, A.; Jossang, P.; Hadi, H. A.; Sevenet, T.; Bodo, B. Horsfiline, An Oxindole Alkaloid From *Horsfieldia superba*. *J. Org. Chem.* **1991**, *56* (23), 6527−6530.
5. Kamisaki, H.; Nanjo, T.; Tsukano, C.; Takemoto, Y. Domino Pd-Catalyzed Heck Cyclization and Bismuth-Catalyzed Hydroamination: Formal Synthesis of Elacomine and Isoelacomine. *Chem. Eur. J.* **2011**, *17* (2), 626−633.
6. Chou, C.-H.; Gong, C.-L.; Chao, C.-C.; Lin, C.-H.; Kwan, C.-Y.; Hsieh, C.-L., et al. Rhynchophylline From *Uncaria rhynchophylla* Functionally Turns Delayed Rectifiers Into A-Type K + Channels. *J. Nat. Prod.* **2009**, *72* (5), 830−834.
7. Stump, C. A.; Bell, I. M.; Bednar, R. A.; Bruno, J. G.; Fay, J. F.; Gallicchio, S. N., et al. The Discovery of Highly Potent CGRP Receptor Antagonists. *Bioorg. Med. Chem. Lett.* **2009**, *19* (1), 214−217.
8. Fensome, A.; Adams, W. R.; Adams, A. L.; Berrodin, T. J.; Cohen, J.; Huselton, C., et al. Design, Synthesis, and SAR of New Pyrrole-Oxindole Progesterone Receptor Modulators Leading to 5-(7-Fluoro-3, 3-dimethyl-2-oxo-2, 3-dihydro-1 H-indol-5-yl)-1-methyl-1 H-pyrrole-2-Carbonitrile (WAY-255348). *J. Med. Chem.* **2008**, *51* (6), 1861−1873.
9. Schreiber, S. L. Target-Oriented and Diversity-Oriented Organic Synthesis in Drug Discovery. *Science* **2000**, *287* (5460), 1964−1969.

CHAPTER 2

(−)-Affinisine

Gardneria sarpagine related (−)-affinisine spirooxindole (Fig. 2.1) belongs to the "alkaloid I" type and was first isolated by Kam et al. from the leaf extract of *Alstonia angustifolia* var. latifolia and its structure was consigned in 2004. These alkaloids possess short-lived in vivo ganglionic transmission inhibitory activity both in rabbits and rats.[1,2]

FIGURE 2.1 Structure of (−)-affinisine.

2.1 Fonseca's first stereospecific total synthesis of (−)-affinisine (2014)

See Scheme 2.1.

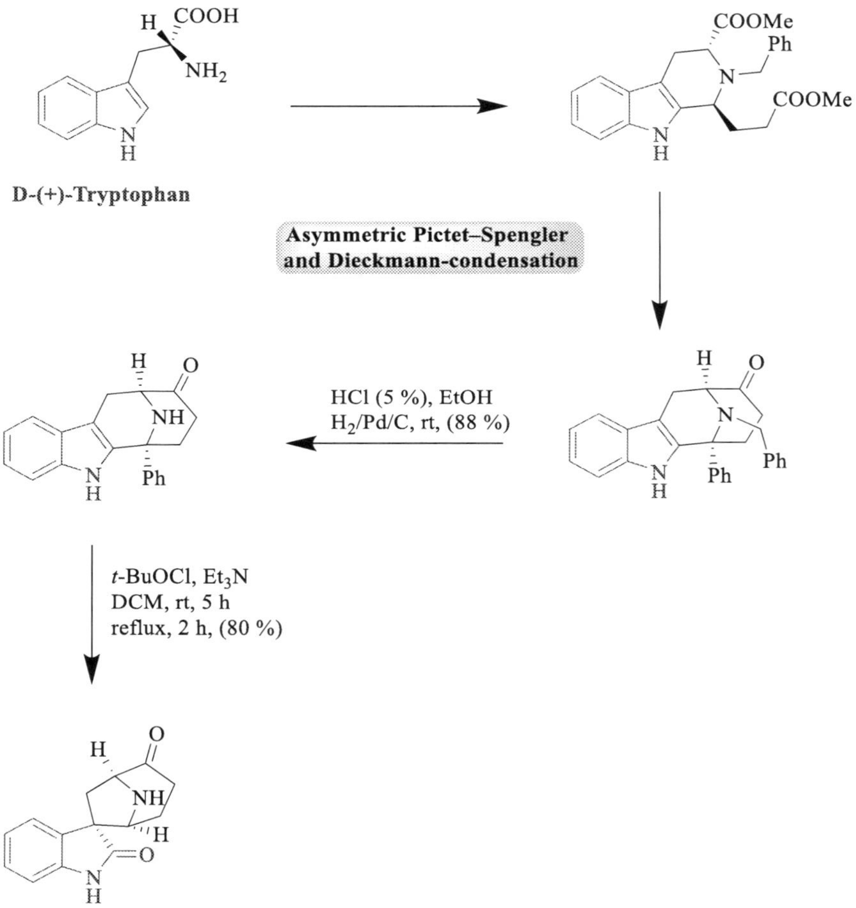

SCHEME 2.1 Fonseca's first stereospecific total synthesis of (−)-affinisine.[2]

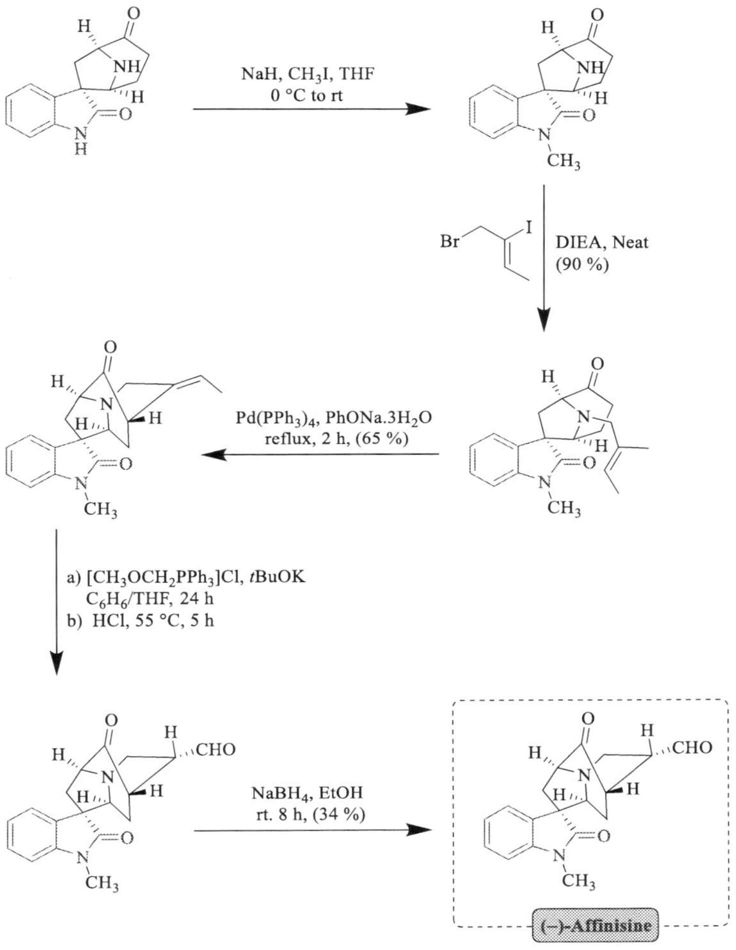

2.1 Fonseca's first stereospecific total synthesis of (−)-affinisine (2014)

SCHEME 2.1 (Continued).

Key features

- Asymmetric Pictet−Spengler/Dieckmann-condensation sequence applied on readily available D-(+)-tryptophan precursor to accomplish tetracyclic ketone
- Stereospecific construction of spirocyclic oxindoles on to the produced tetracyclic ketone
- Synthesis of N_b-allyl oxindole
- Conversion of N_b-allyl oxindole intermediate into (−)-affinisine spirooxindole

8

2.2 Improved total synthesis of (−)-affinisine (2017)

See Scheme 2.2.

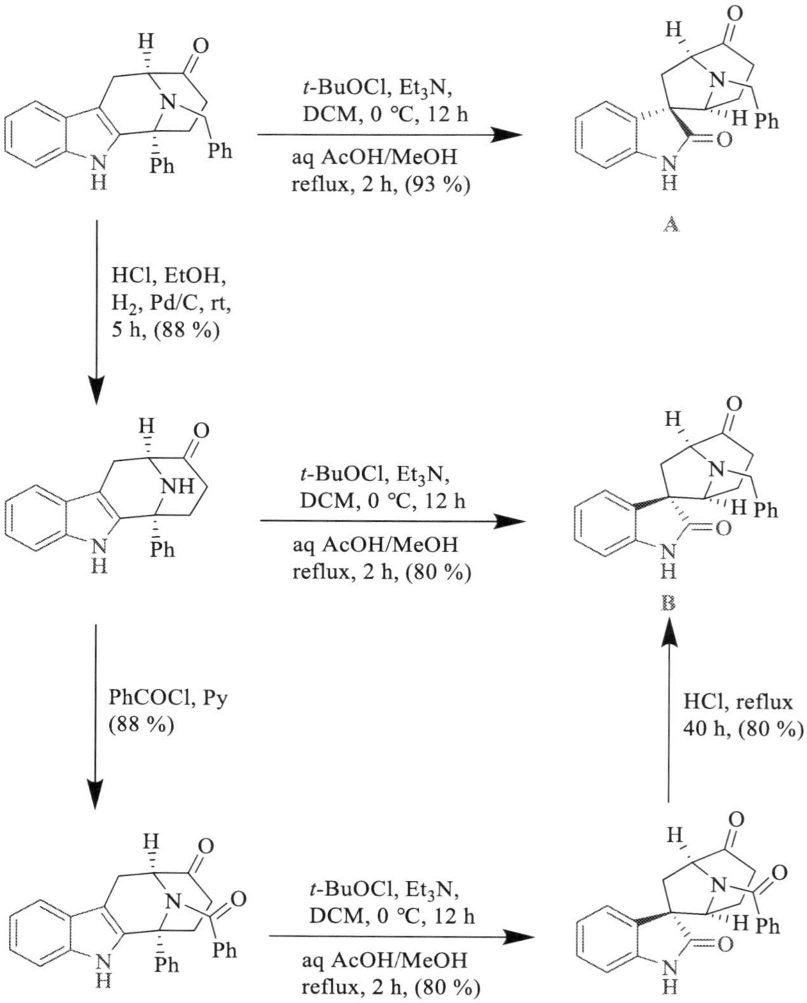

SCHEME 2.2 Fonseca's improved total synthesis of (−)-affinisine.[1]

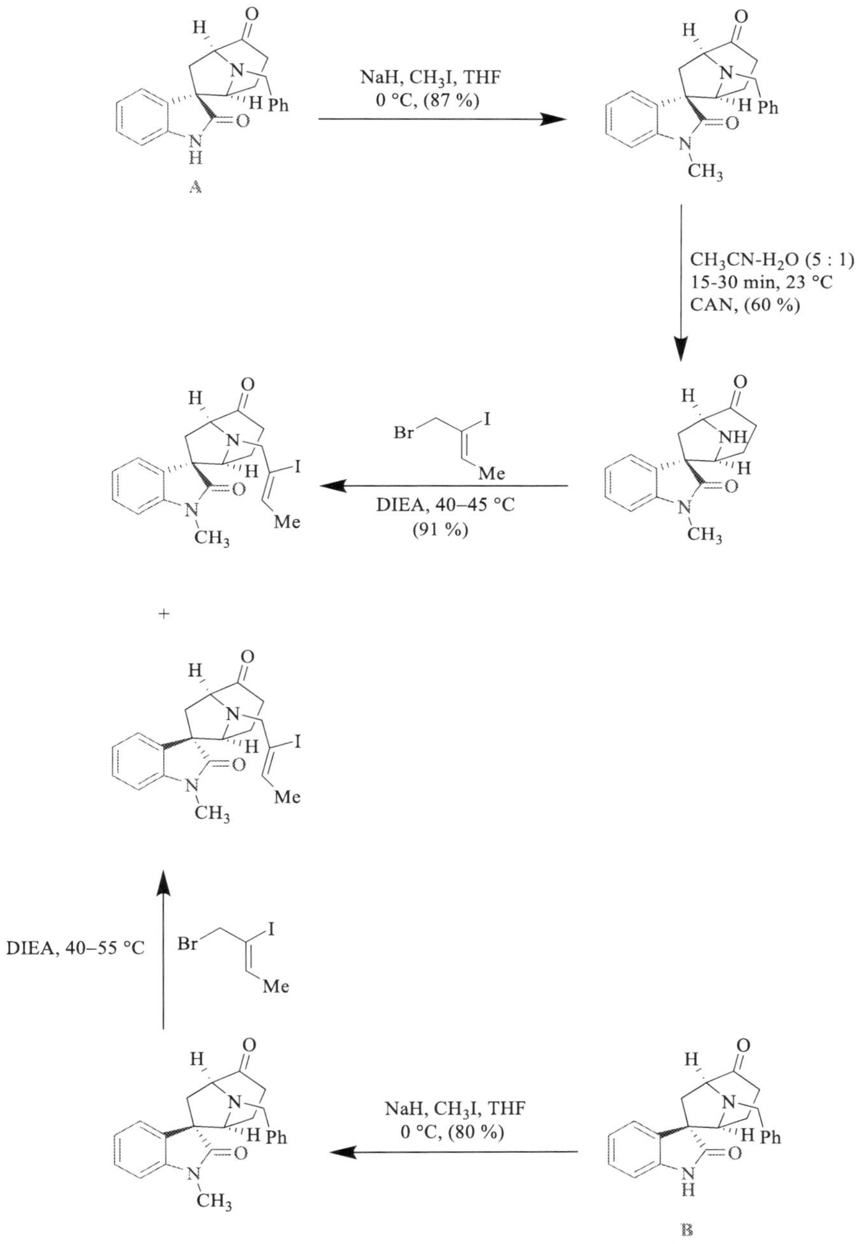

2.2 Improved total synthesis of (−)-affinisine (2017)

SCHEME 2.2 (Continued).

Indole Alkaloids

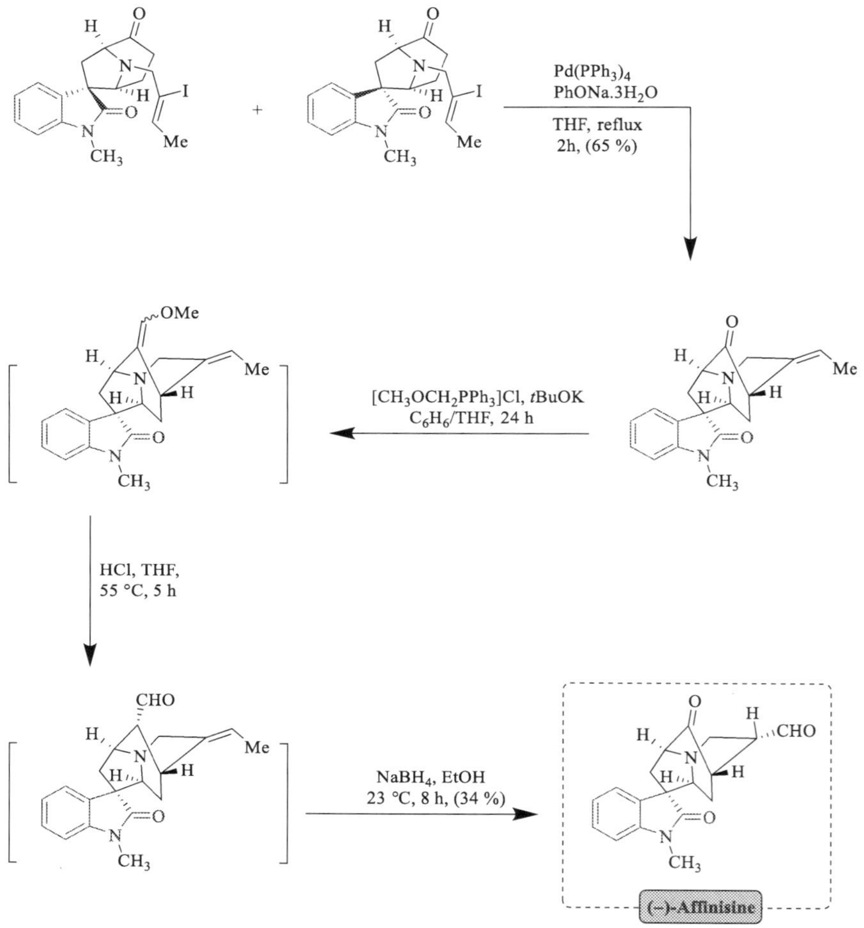

10

2. (−)-Affinisine

SCHEME 2.2 (Continued).

Key features

- Readily available D-(+)-tryptophan was used as precursor.
- Diastereospecific spiro center was generated via *tert*-butyl hypochlorite-promoted oxidative rearrangement of a chiral tetrahydro-β-carboline derivative. This key branching point determined the spatial configuration at the C-7 spiro center to be entirely 7R or 7S.
- Asymmetric Pictet–Spengler reaction and Dieckmann cyclization were the other key stereospecific processes.

References

1. Fonseca, G. O.; Wang, Z.-J.; Namjoshi, O. A.; Deschamps, J. R.; Cook, J. M. First Stereospecific Total Synthesis of (−)-Affinisine Oxindole as well as Facile Entry Into the C (7)-Diastereomeric Chitosenine Stereochemistry. *Tetrahedron Lett.* **2015,** *56* (23), 3052−3056.
2. Stephen, M. R.; Rahman, M. T.; Tiruveedhula, V. P. B.; Fonseca, G. O.; Deschamps, J. R.; Cook, J. M. Concise Total Synthesis of (−)-Affinisine Oxindole, (+)-Isoalstonisine, (+)-Alstofoline, (−)-Macrogentine, (+)-Na-Demethylalstonisine, (−)-Alstonoxine A, and (+)-Alstonisine. *Chem. Eur. J.* **2017,** *23* (62), 15805−15819.

CHAPTER 3

(+)-Austamide

Austamide (Fig. 3.1) is a toxic metabolite isolated from *Aspergillus ustus* in 1971 by Steyn.[1] Its structure elucidation was done in 1973 by the same research group.[2]

From the synthetic point of view the presence of a delicate indoxyl chromophore and two labile enamide functionalities make it difficult to synthesize.[3]

Biosynthesis of austamide is thought to proceed through the intermediacy of deoxybrevianamide E. Thus "reverse" prenylation of brevianamide F (*cyclo*-L-Trp-L-Pro) produces deoxybrevianamide E, which undergoes diverse oxidative cyclizations leading to austamide in *A. ustus* and brevianamides A and B in *Penicillium* sp. The intermediacy of deoxybrevianamide E in the process has been confirmed by incorporating tritium labeled deoxybrevianamide E in cultures of *Penicillium brevicompactum* as reported by Williams et al.[4,5] (Fig. 3.2).

FIGURE 3.1 Structure of (+)-austamide.

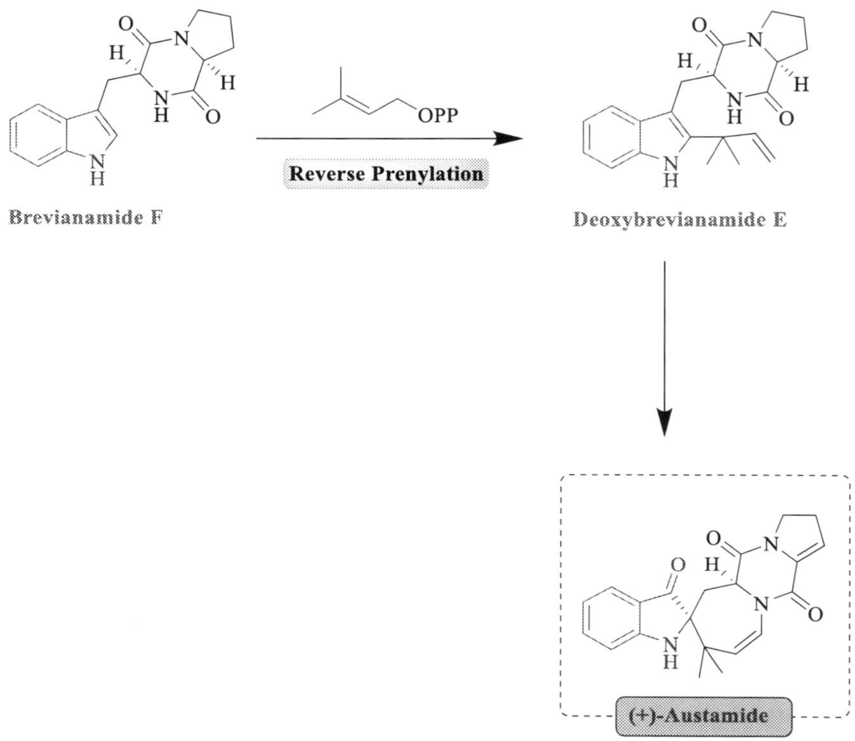

14

3. (+)-Austamide

Brevianamide F

Reverse Prenylation

Deoxybrevianamide E

(+)-Austamide

FIGURE 3.2 Biosynthesis of austamide.

3.1 Hutchison's total synthesis of (±)-austamide (1979)

See Scheme 3.1.

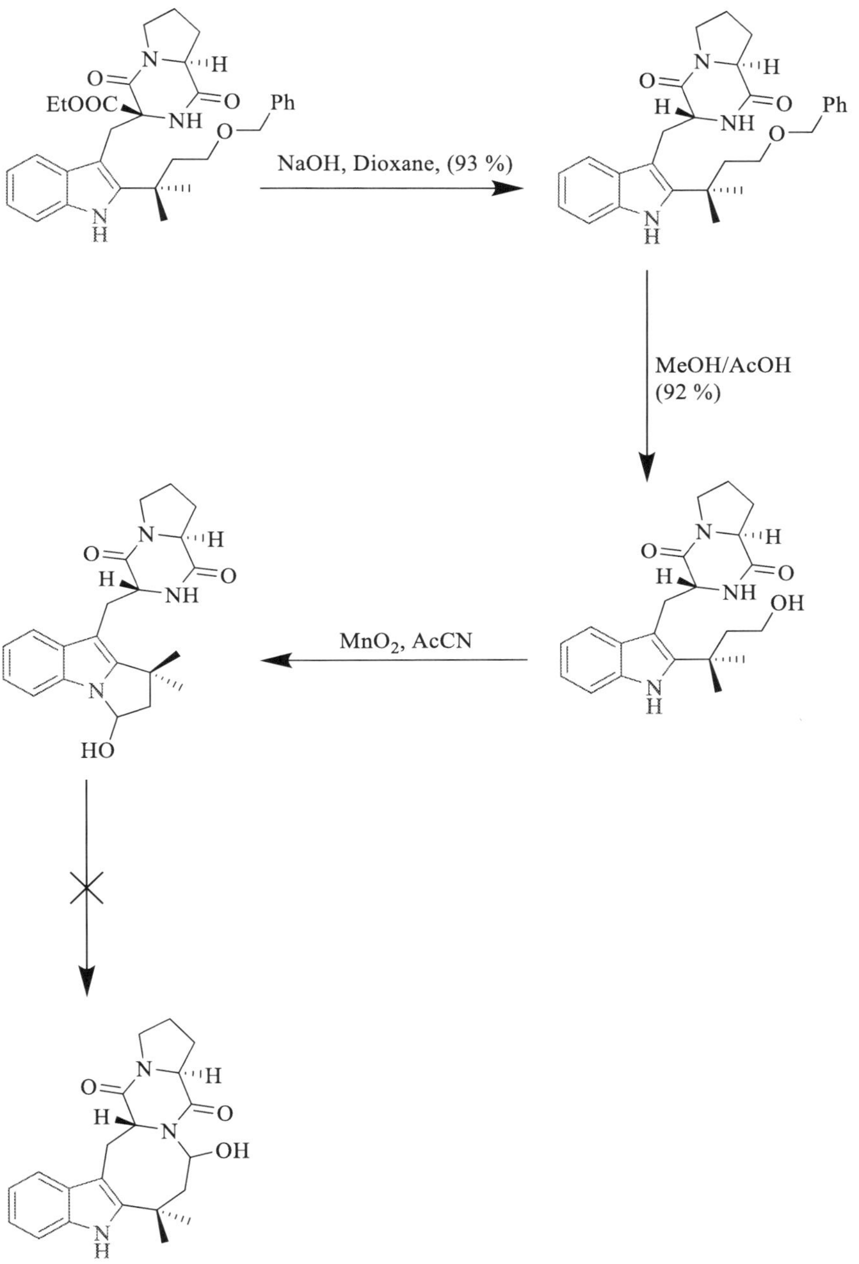

3.1 Hutchison's total synthesis of (±)-austamide (1979) 15

SCHEME 3.1   Hutchison's total synthesis of (±)-austamide.[6]

Indole Alkaloids

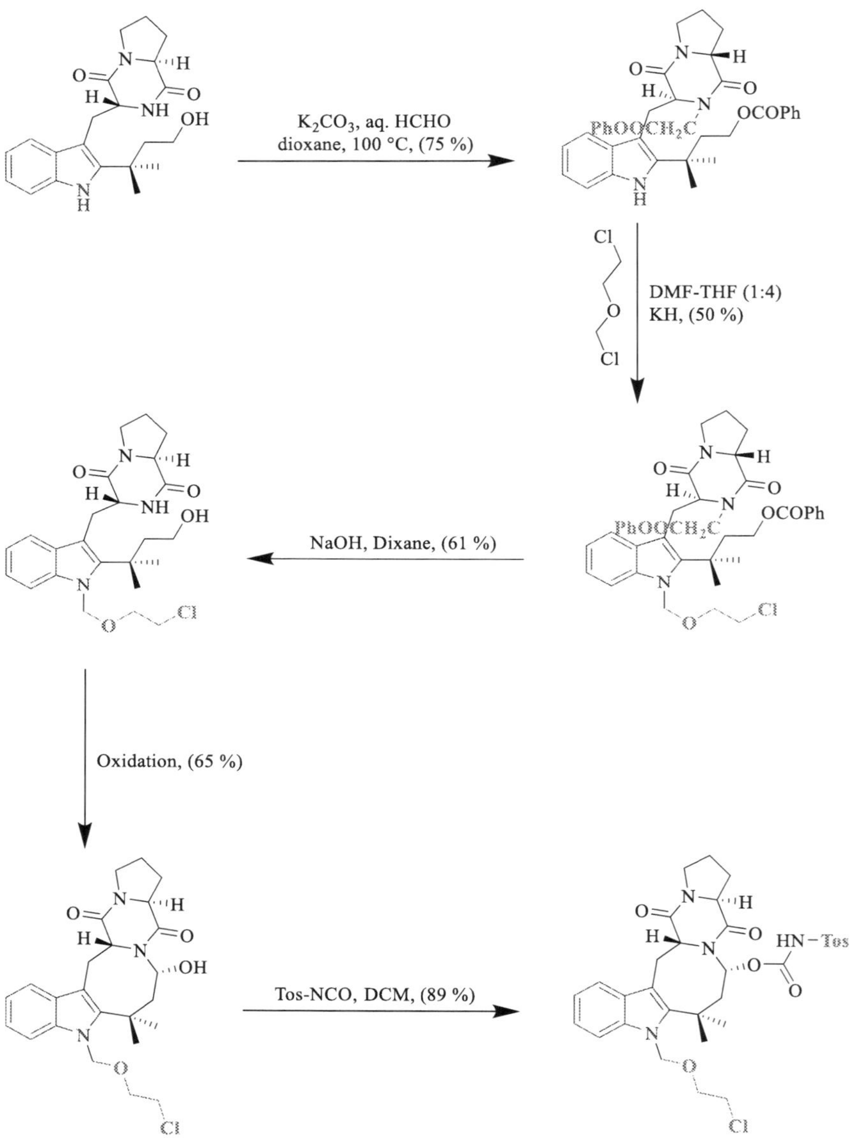

3. (+)-Austamide

SCHEME 3.1 (Continued).

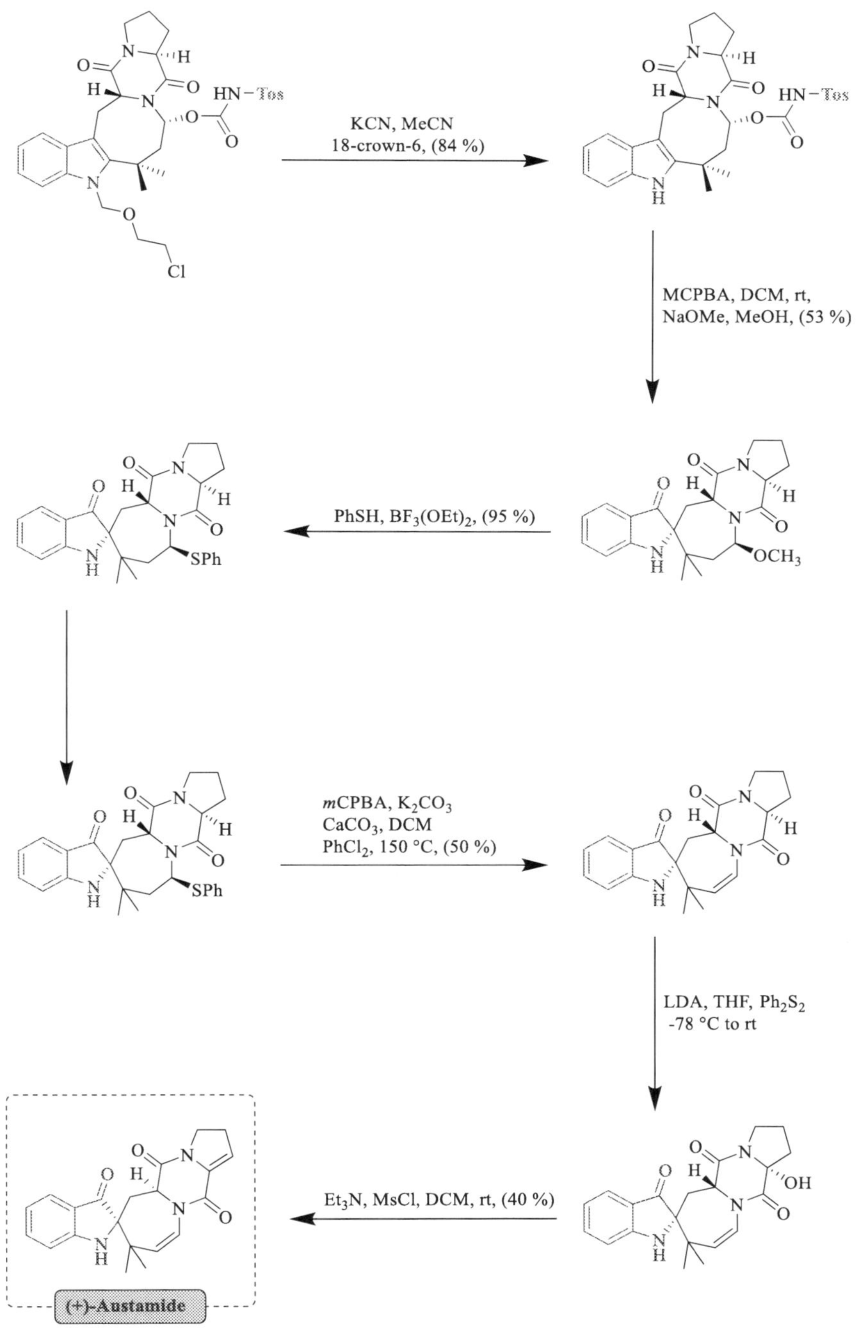

3.1 Hutchison's total synthesis of (±)-austamide (1979)

SCHEME 3.1 (Continued).

18

3. (+)-Austamide

Key features

- Hydrolysis and decarboxylation of diketopiperazine ester
- Hydrogenolysis
- Hydroxymethylation followed by benzoylation
- N-protection via alkylation with 2-chloroethyl chloromethyl ether
- Hydrolysis followed by oxidation to yield *N*-acyl-*N,O*-hemiacetal
- Protection to yield tosyl carbamate followed by deprotection of 2-chloroethoxymethylene group
- Oxidation and rearrangement to yield austamide

3.2 Corey's total synthesis of (+)-austamide (2000)

See Scheme 3.2.

Indole Alkaloids

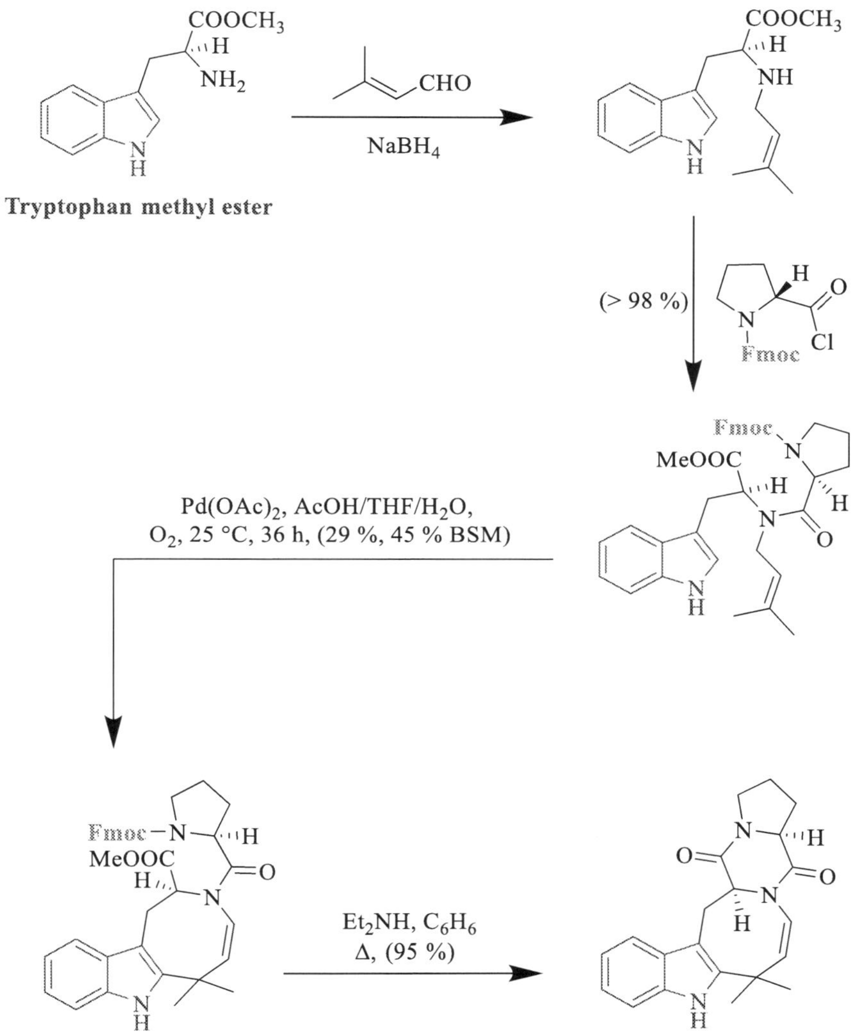

3.2 Corey's total synthesis of (+)-austamide (2000)

19

SCHEME 3.2 Corey's total synthesis of (+)-austamide.[7]

Indole Alkaloids

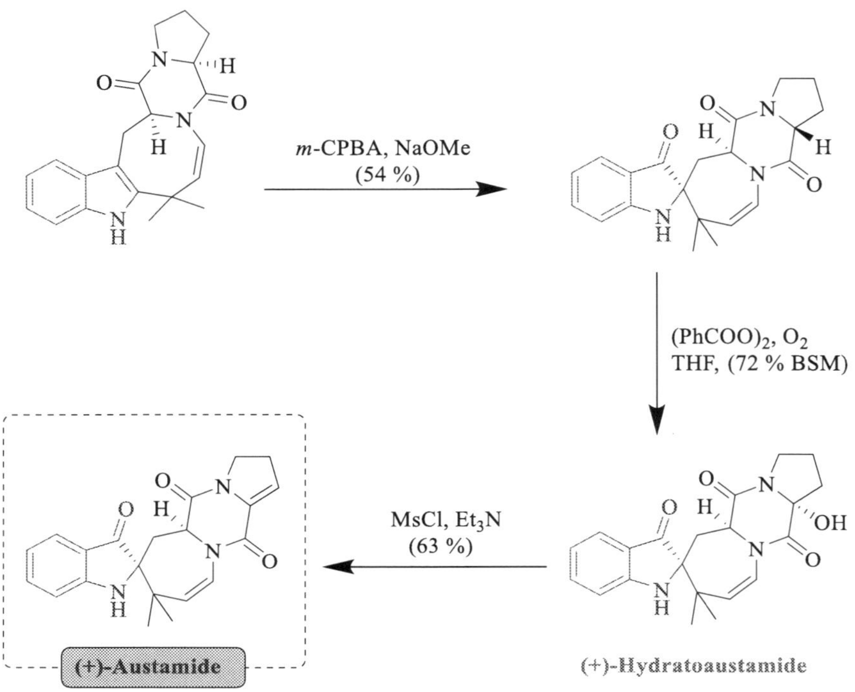

20

3. (+)-Austamide

SCHEME 3.2 (Continued).

Key features

- Synthesis of (+)-austamide starting from readily available (*S*)-tryptophan methyl ester
- Conversion of (*S*)-tryptophan methyl ester into the Schiff base with 3-methyl-2-butenal and subsequent reduction
- Reaction of obtained reduced product with Fmoc (*S*)-prolyl chloride to produce coupled amide
- Cyclization to generate an eight-membered ring
- Generation of spiro center and obtainment of final alkaloid

References

1. Steyn, P. Austamide, A New Toxic Metabolite From *Aspergillus ustus*. *Tetrahedron Lett.* **1971,** *12* (36), 3331–3334.
2. Steyn, P. The Structures of Five Diketopiperazines From *Aspergillus ustus*. *Tetrahedron* **1973,** *29* (1), 107–120.
3. Stocking, E. M.; Williams, R. M.; Sanz-Cervera, J. F. Reverse Prenyl Transferases Exhibit Poor Facial Discrimination in the Biosynthesis of Paraherquamide A, Brevianamide A, and Austamide. *J. Am. Chem. Soc.* **2000,** *122* (38), 9089–9098.

Indole Alkaloids

References

4. Sanz-Cervera, J. F.; Glinka, T.; Williams, R. M. Biosynthesis of Brevianamides A and B: In Search of the Biosynthetic Diels-Alder Construction. *Tetrahedron* **1993,** *49* (38), 8471–8482.
5. Williams, R. M.; Stocking, E. M.; Sanz-Cervera, J. F. *Biosynthesis of Prenylated Alkaloids Derived From Tryptophan. Biosynthesis;* Springer: Berlin, Heidelberg, **2000,** 97–173.
6. Hutchison, A. J.; Kishi, Y. Stereospecific Total Synthesis of DL-Austamide. *J. Am. Chem. Soc.* **1979,** *101* (22), 6786–6788.
7. Baran, P. S.; Corey, E. A Short Synthetic Route to (+)-Austamide, (+)-Deoxyisoaustamide, and (+)-Hydratoaustamide From a Common Precursor by a Novel Palladium-Mediated Indole→Dihydroindoloazocine Cyclization. *J. Am. Chem. Soc.* **2002,** *124* (27), 7904–7905.

CHAPTER 4

Brevianamide A and B

Brevianamides A and B belong to bicyclo[2.2.2]diazaoctane alkaloids exhibiting dioxopiperazine-type structures (Fig. 4.1). Brevianamide A was first isolated in 1969 by Birch and his research fellows from *Penicillium brevicompactum* fungus as a metabolite.[1] Later on, brevianamide B was also isolated from the same fungus.

These molecules are of profound interest to chemists as well as biologists due to synthetically intimidating structures, diverse biosynthetic origins, and significant biological activities.

Brevianamide A is a potent antifeedant molecule against the larvae of the insect pests *Heliothis virescens* (tobacco budworm) and *Spodoptera frugiperda* (fall armyworm).[2,3] Brevianamide molecules are members of mycotoxins; curious and still increasing small class of alkaloids that have just been linked with marcfortine and paraherquamides exhibiting potent antiparasitic properties.[4]

Brevianamide A is the most abundant metabolite, and its biosynthesis occurs in nature via enzyme-free Diels−Alder reaction. Contrarily, there are very few cases where nature uses Diels−Alderase for biosynthesis. As both of these molecules and other members of bicyclo[2.2.2]diazaoctane alkaloids are isolated in an enantiopure form, the reaction

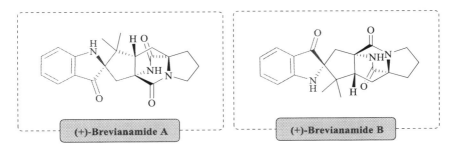

FIGURE 4.1 Structures of (+)-Brevianamides A and B.

is likely to be enantio- and stereocontrolled, suggesting the possible use of an enzyme.

According to the first proposed biosynthetic hypothesis by Birch, biosynthesis is initiated by the production of brevianamide F molecule via condensation of *L*-tryptophan and *L*-proline. This brevianamide F is then converted to deoxybrevianamide E via reverse prenylation which, in turn, undergoes proposed intramolecular Diels−Alder reaction giving a racemic mixture of products. Racemic intermediates are then oxidized with subsequent pinacol rearrangement to give brevianamides A and B both (Fig. 4.2).[5] In vitro calculations of energy transition states of the products showed that the product corresponding to brevinamide A

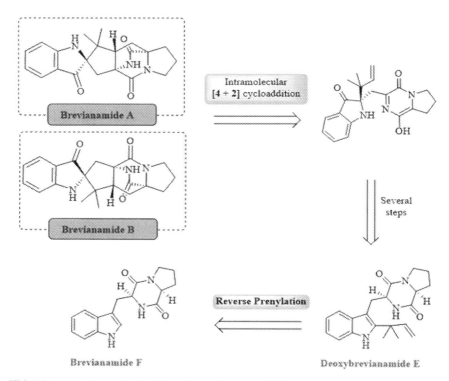

FIGURE 4.2   Biosynthesis of brevianamides A and B.[5,6]

had a lower energy transition state than that of brevinamide B reasoning the brevianamide A as a major metabolite in nature.[6]

Talking about chemical synthesis, it is still an obscure target for brevianamide A; even after five research decades, only one total synthesis is reported.[7] On the other hand, several synthetic methods for brevianamide B (minor diastereomer) have been reported so far including two total syntheses by Williams[8] and Lawrence.[9]

4.1 Total syntheses of brevianamides A and B

4.1.1 Williams' asymmetric total synthesis of brevianamide B (1990)

See Scheme 4.1.

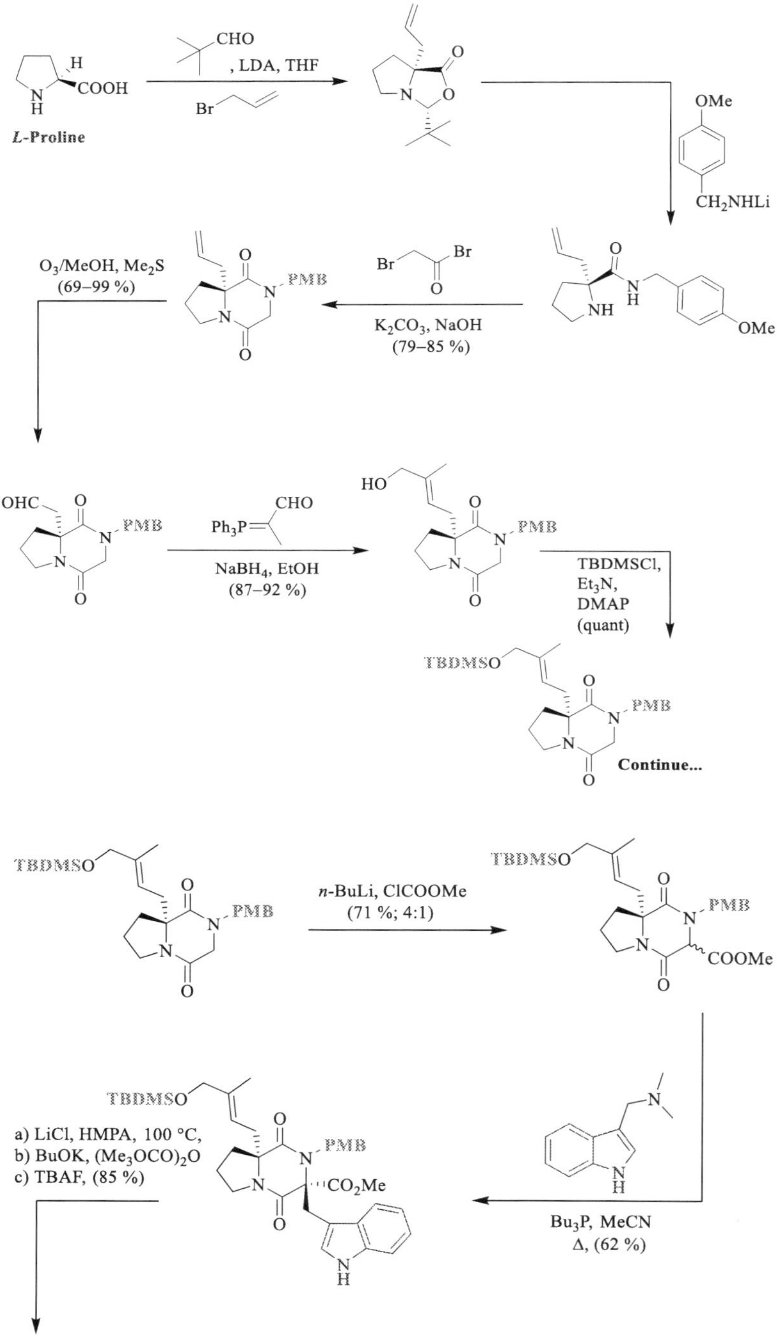

4. Brevianamide A and B

SCHEME 4.1 Williams' asymmetric total synthesis of (−)-brevianamide B.[8]

Indole Alkaloids

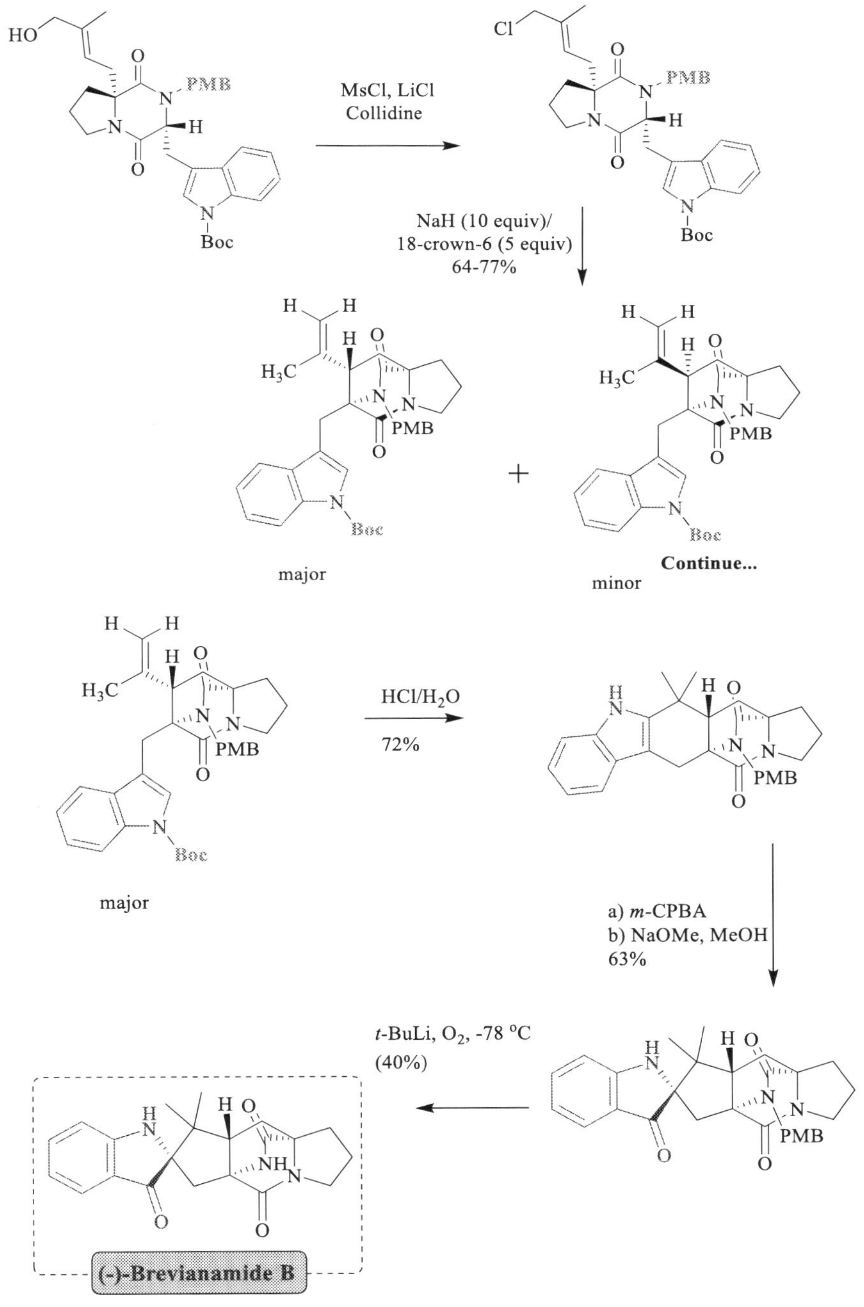

4.1 Total syntheses of brevianamides A and B

27

SCHEME 4.1 (Continued).

Indole Alkaloids

28
4. Brevianamide A and B

Key features

- Intramolecular S_N2' cyclization in a stereocontrolled way to build the central bicyclo[2.2.2] scaffold
- Formation of pivaldehyde acetal from *L*-proline
- Ring closure to afford enantiomerically pure piperazinedione
- Ozonolysis to form optically pure aldehyde
- Spiro cyclization

4.1.2 Lawrence's total synthesis of brevianamides A and B (2019, 2020)

See Scheme 4.2.

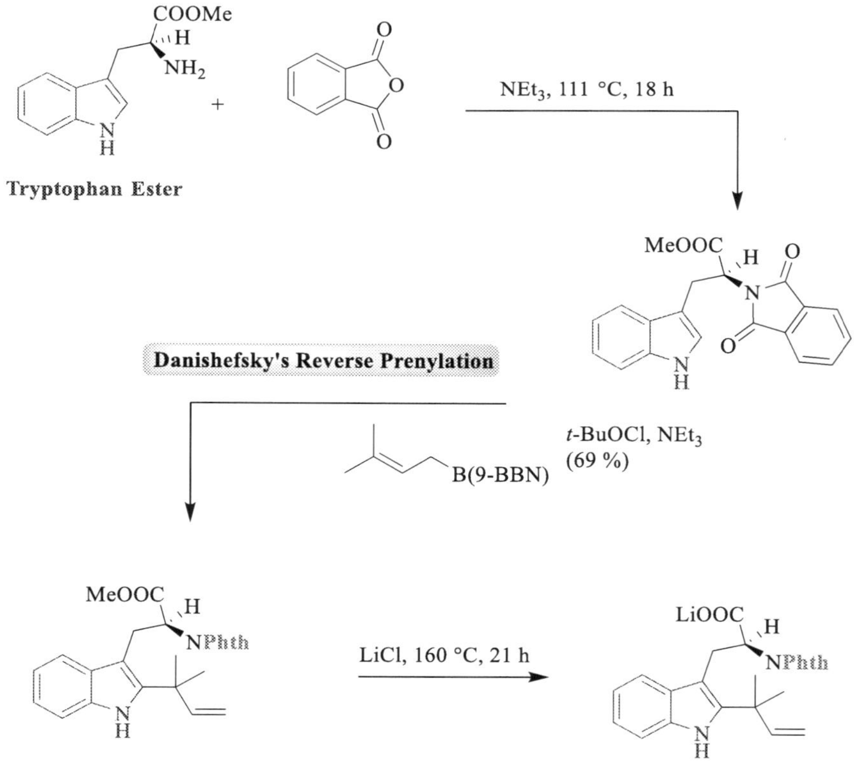

SCHEME 4.2 Lawrence's total synthesis of brevianamides A and B.[7,9]

Indole Alkaloids

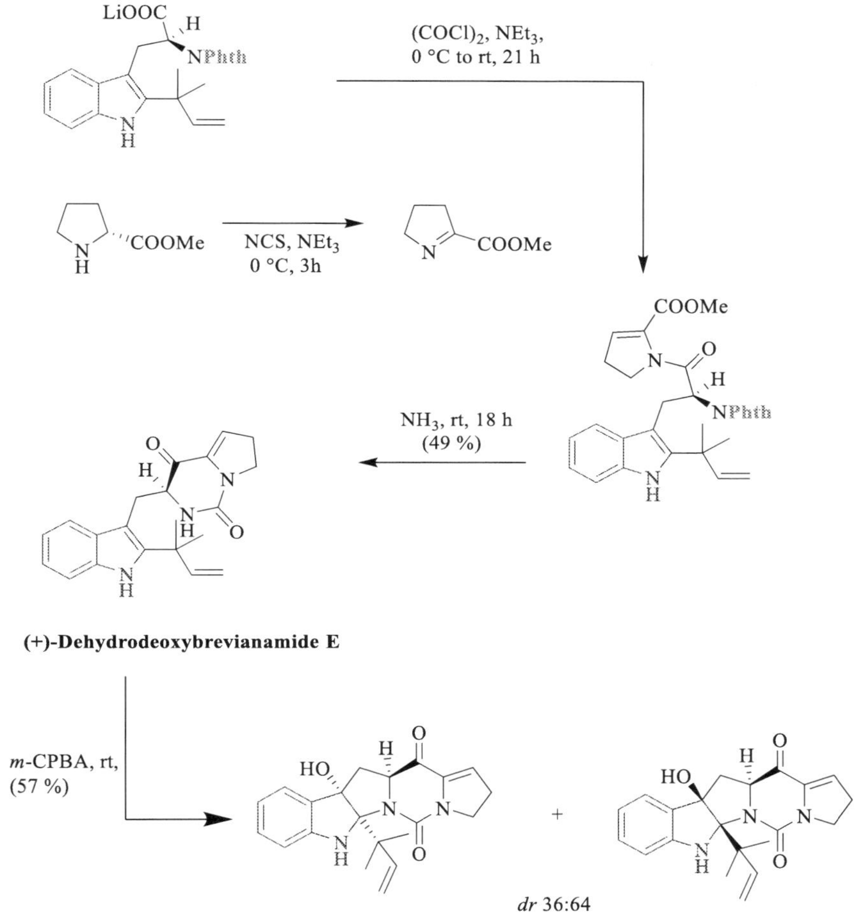

4.1 Total syntheses of brevianamides A and B

29

(+)-Dehydrodeoxybrevianamide E

dr 36:64

SCHEME 4.2 (Continued).

Indole Alkaloids

30   4. Brevianamide A and B

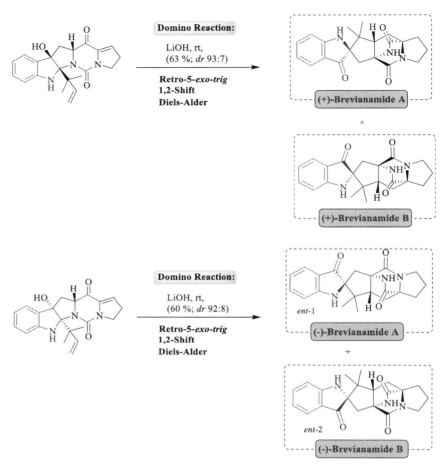

SCHEME 4.2 (Continued).

4.1 Total syntheses of brevianamides A and B

Key features

- Synthesis of (+)-brevianamide A was completed in seven steps with 7.2% overall yield.
- Key bioinspired one-step cascade transformation of linearly fused dehydrobrevianamide E into bridged-spiro-fused brevianamide A.
- Danishefsky's reverse prenylation.
- One-pot formation of *N*-acyl enamine via acylation of lithium carboxylate and imine acylation reaction with dehydroproline.
- Domino reaction in end game to afford brevianamides A and B and their enantiomers, respectively.

4.1.3 Scheerer's formal synthesis of brevianamide B (2016)

See Scheme 4.3.

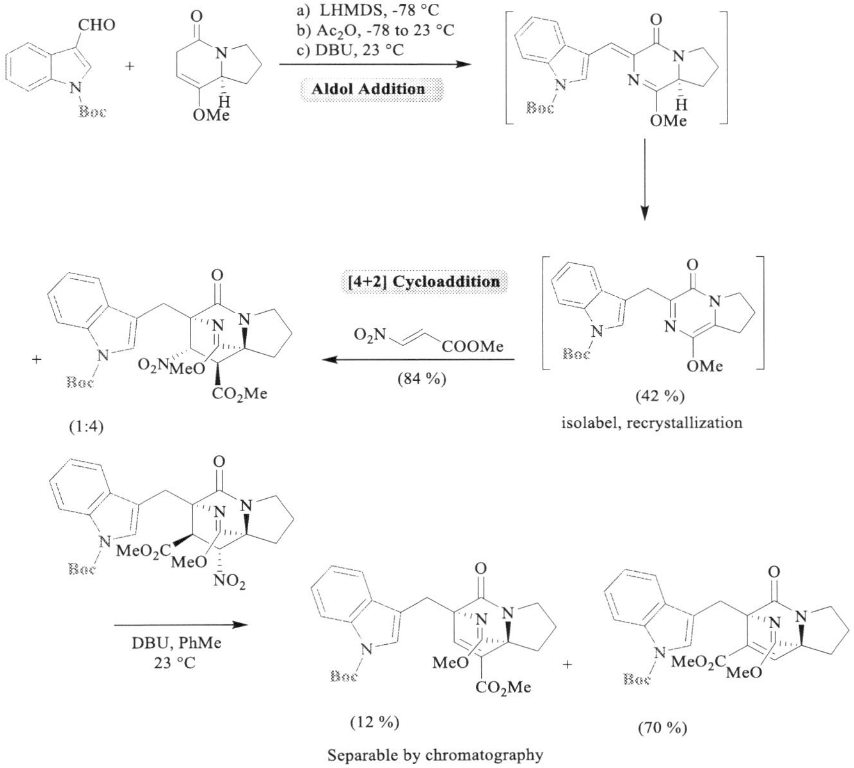

SCHEME 4.3 Scheerer's formal synthesis of brevianamide B.[10]

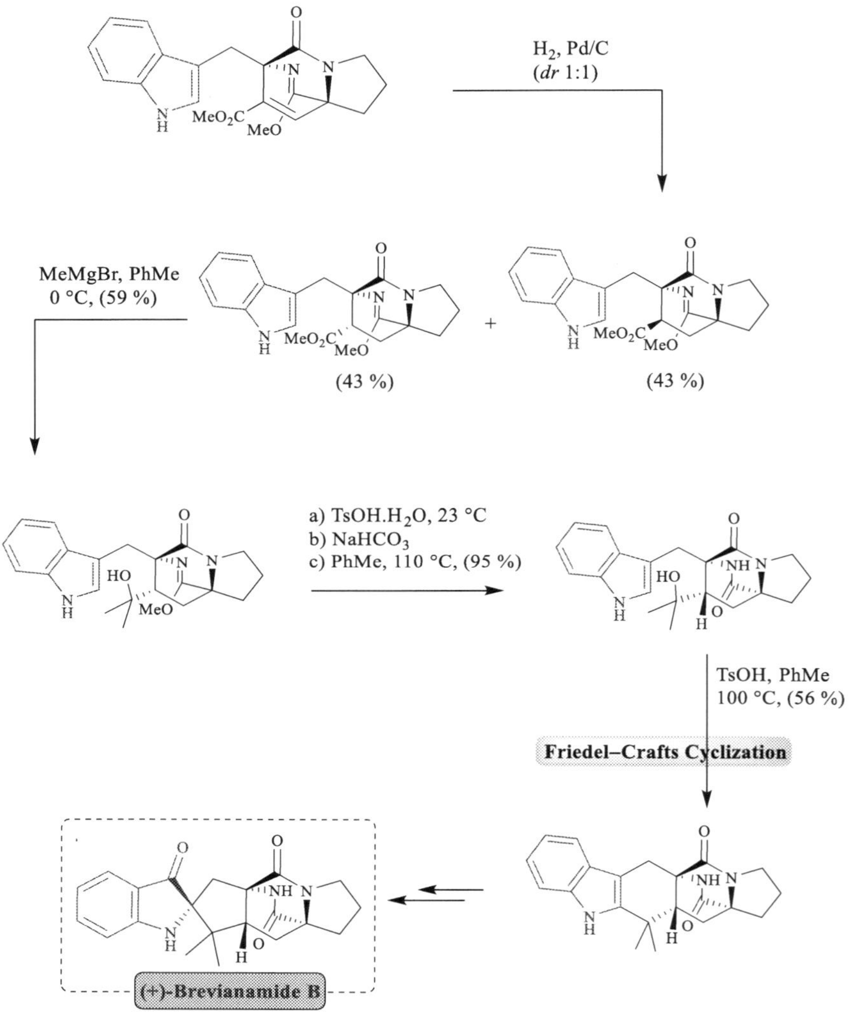

4. Brevianamide A and B

SCHEME 4.3 (Continued).

Key features

- Synthesis of pyrazinone from proline-derived diketopiperazine via enolization followed by aldol addition of *N*-protected indole aldehyde
- Cycloaddition of synthesized pyrazinone with methyl 2-nitroacrylate dienophile
- Removal of nitro group

- Methylation of ester
- Friedel−Crafts cyclization

References

1. Birch, A.; Wright, J. The Brevianamides: A New Class of Fungal Alkaloid. *J. Chem. Soc. D, Chem. Commun.* **1969,** *12,* 644b−645b.
2. Paterson, R.; Simmonds, M.; Kemmelmeier, C.; Blaney, W. Effects of Brevianamide A, Its Photolysis Product Brevianamide D, and Ochratoxin A From Two *Penicillium* Strains on the Insect Pests *Spodoptera frugiperda* and *Heliothis virescens. Mycol. Res.* **1990,** *94* (4), 538−542.
3. Klas, K. R.; Kato, H.; Frisvad, J. C.; Yu, F.; Newmister, S. A.; Fraley, A. E.; Sherman, D. H.; Tsukamoto, S.; Williams, R. M. Structural and Stereochemical Diversity in Prenylated Indole Alkaloids Containing the Bicyclo[2.2. 2]diazaoctane Ring System From Marine and Terrestrial Fungi. *Nat. Prod. Rep.* **2018,** *35* (6), 532−558.
4. Yamazaki, M.; Okuyama, E.; Kobayashi, M.; Inoue, H. The Structure of Paraherquamide, A Toxic Metabolite From *Penicillium paraherquei. Tetrahedron Lett.* **1981,** *22* (2), 135−136.
5. Williams, R. M.; Cox, R. J. Paraherquamides, Brevianamides, and Asperparalines: Laboratory Synthesis and Biosynthesis. An Interim Report. *Acc. Chem. Res.* **2003,** *36* (2), 127−139.
6. Domingo, L. R.; Sanz-Cervera, J. F.; Williams, R. M.; Picher, M. T.; Marco, J. A. Biosynthesis of the Brevianamides. An Ab Initio Study of the Biosynthetic Intramolecular Diels−Alder Cycloaddition. *J. Org. Chem.* **1997,** *62* (6), 1662−1667.
7. Godfrey, R. C.; Green, N. J.; Nichol, G. S.; Lawrence, A. L. Total Synthesis of Brevianamide A. *Nat. Chem.* **2020,** *12* (7), 615−619.
8. Williams, R. M.; Glinka, T.; Kwast, E.; Coffman, H.; Stille, J. K. Asymmetric, Stereocontrolled Total Synthesis of (−)-Brevianamide B. *J. Am. Chem. Soc.* **1990,** *112* (2), 808−821.
9. Godfrey, R.; Green, N.; Nichol, G.; Lawrence, A. Chemical Synthesis of (+)-Brevianamide A Supports a Diels−Alderase-Free Biosynthesis. *ChemRxiv Prepr.* **2019**.
10. Robins, J. G.; Kim, K. J.; Chinn, A. J.; Woo, J. S.; Scheerer, J. R. Intermolecular Diels−Alder Cycloaddition for the Construction of Bicyclo[2.2. 2]diazaoctane Structures: Formal Synthesis of Brevianamide B and Premalbrancheamide. *J. Org. Chem.* **2016,** *81* (6), 2293−2301.

CHAPTER 5

Citrinadin A and B

Citrinadins A and B exhibiting a spiro-fused oxindole with cyclopenta[*b*]quinolizidine core differ only by C14 ester that is present only in citrinadin A. The structures of molecules are capped with a rare *N,N*-dimethylvaline ester on one end, while a unique epoxycarbonyl side chain is present on the other[1] (Fig. 5.1).

Citrinadin A was isolated by Kobayashi and coworkers in 2004 as a novel secondary metabolite of the marine fungus *Penicillium citrinum* which was cultured from the marine red alga *Actinotrichia fragilis*. This red alga was collected from Japan's Okinawa Island. One year after, the same research group isolated citrinadin B.[2]

In addition to their interesting chemical structure, they possess striking biological activities. Citrinadins A and B are cytotoxic against murine leukemia L1210 cells (IC_{50} 6.2 μg/mL for citrinadin A and 10 μg/mL for citrinadin B). In addition, citrinadin A is noted to be active against human epidermoid carcinoma KB cells with 10 μg/mL IC_{50} value.

Kobayashi et al. proposed two possible biosynthetic routes for citrinadin core from L-tryptophan, L-pipecolic acid, and isoprene. The first option is the revisions to a marcfortine-type structure like C12 methylation and loss of carbonyl from bridging amide. The second possibility is the generation of a dipeptide intermediate converting to citrinadins via redox reactions at C18, C14, and C27 (Fig. 5.2).[1]

In lab synthesis of citrinadin, core has been reported by Martin[3] and Sorensen[4] via oxidative rearrangement of an indole. Wood[5] synthesized the core by intermolecular [3 + 2] nitroarone cycloaddition. Total syntheses of citrinadins have also been reported by Martin and Wood research groups later on.

Indole Alkaloids
DOI: https://doi.org/10.1016/B978-0-323-91674-5.00006-7
© 2022 Elsevier Inc. All rights reserved.

5. Citrinadin A and B

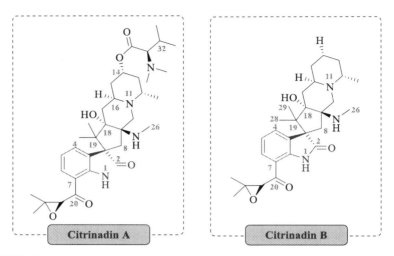

FIGURE 5.1 Structures of citrinadins A and B.

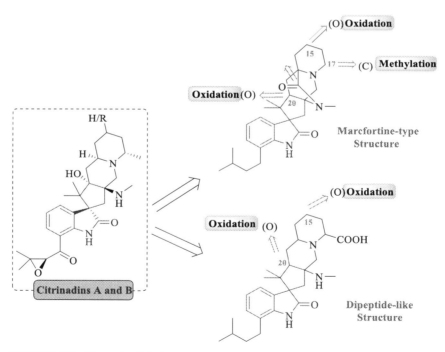

FIGURE 5.2 Two plausible biogenetic precursors in the biosynthetic pathway of citrinadin core proposed by Kobayashi et al.[1]

5.1 Syntheses of citrinadins A and B spirooxindole core

5.1.1 Martin's synthesis of citrinadin A spirooxindole core (2007)

See Scheme 5.1.

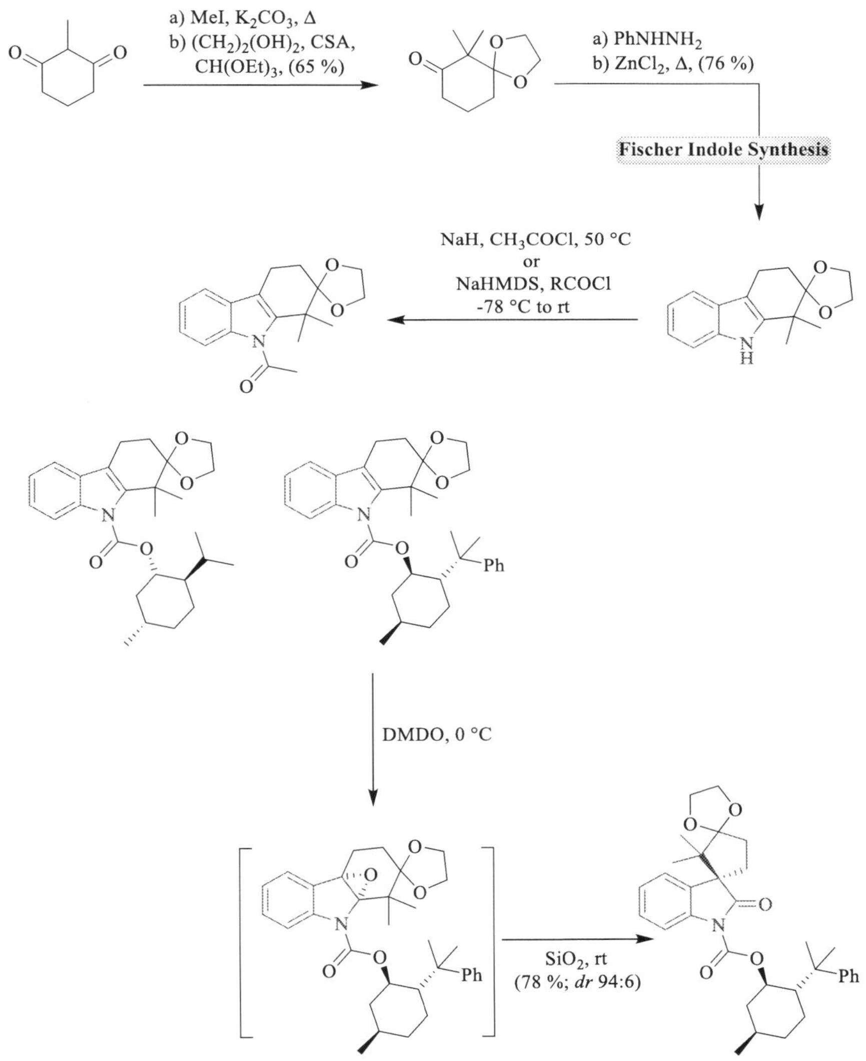

SCHEME 5.1 Martin's formal synthesis of citrinadin A spirooxindole core.[3]

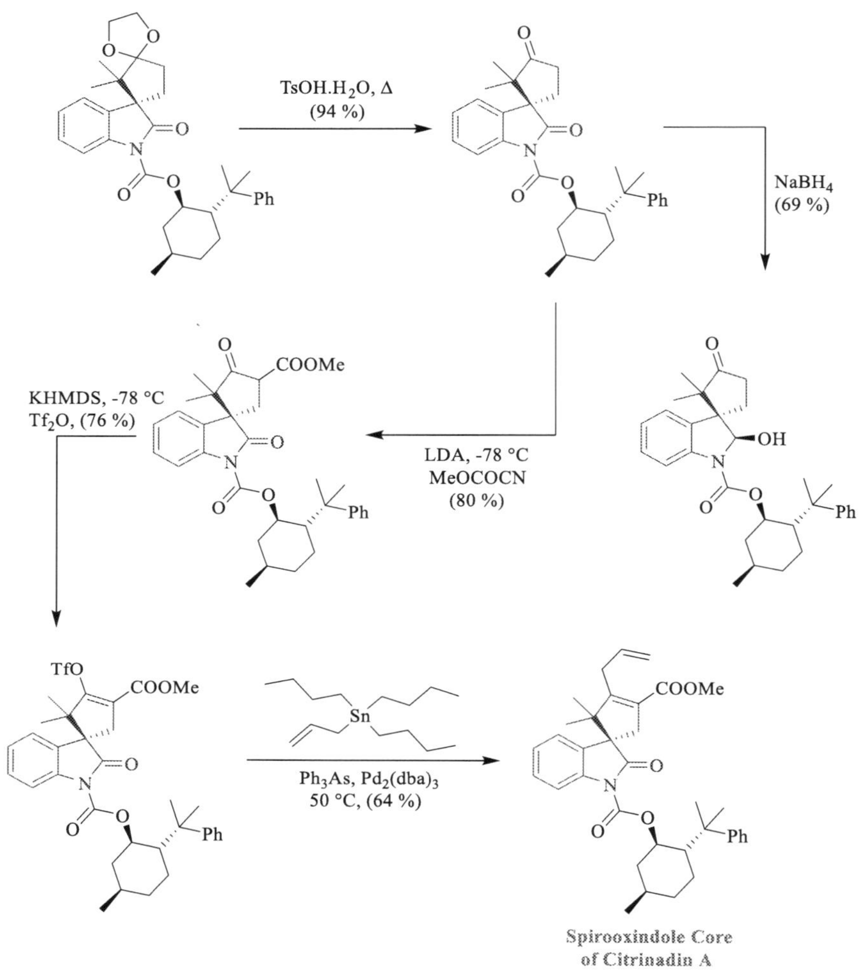

5. Citrinadin A and B

SCHEME 5.1 (Continued).

Key features

- Synthesis of citrinadin A spirooxindole core architecture
- Fischer-indole synthesis from ketal
- *N*-Acylation of prepared indole derivative
- Diastereoselective DMDO-mediated oxidative rearrangements of *N*-acyl derivatives
- Key step involves employment of an 8-phenylmenthol chiral auxiliary on nitrogen of indole

5.1 Syntheses of citrinadins A and B spirooxindole core

39

5.1.2 Sorensen's synthesis of citrinadin B core architecture (2011)

See Scheme 5.2.

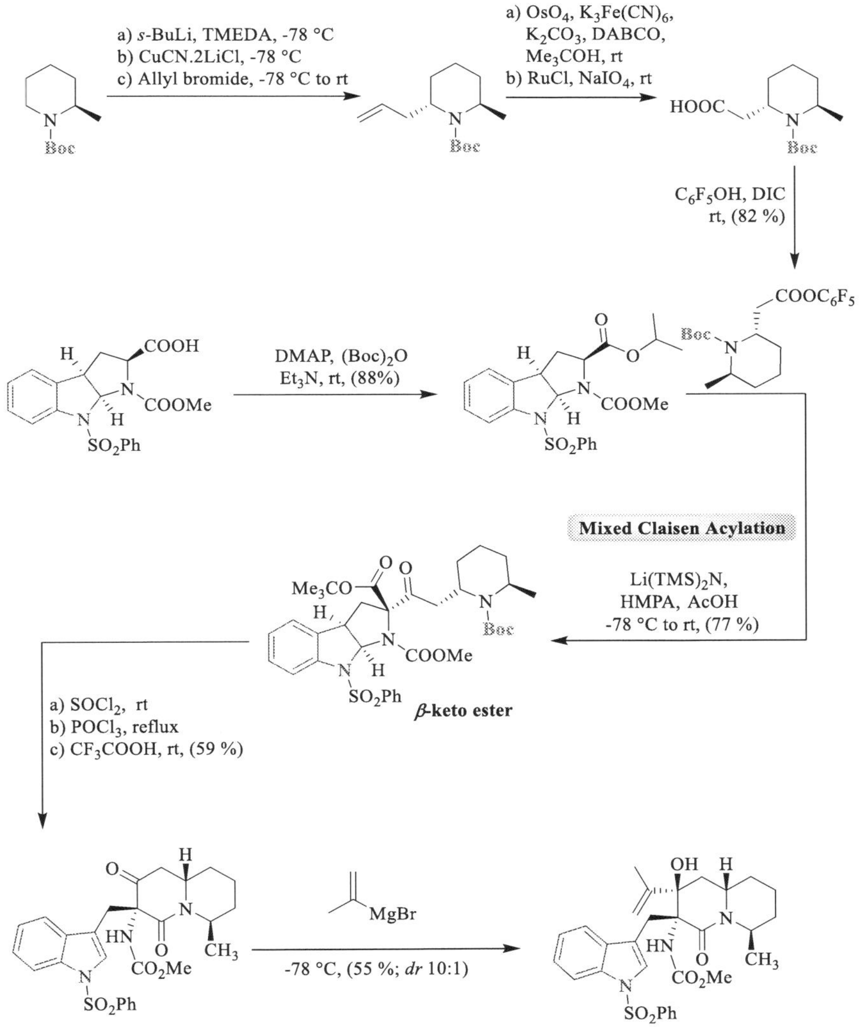

SCHEME 5.2 Sorensen's synthesis of citrinadin B core architecture.[4]

Indole Alkaloids

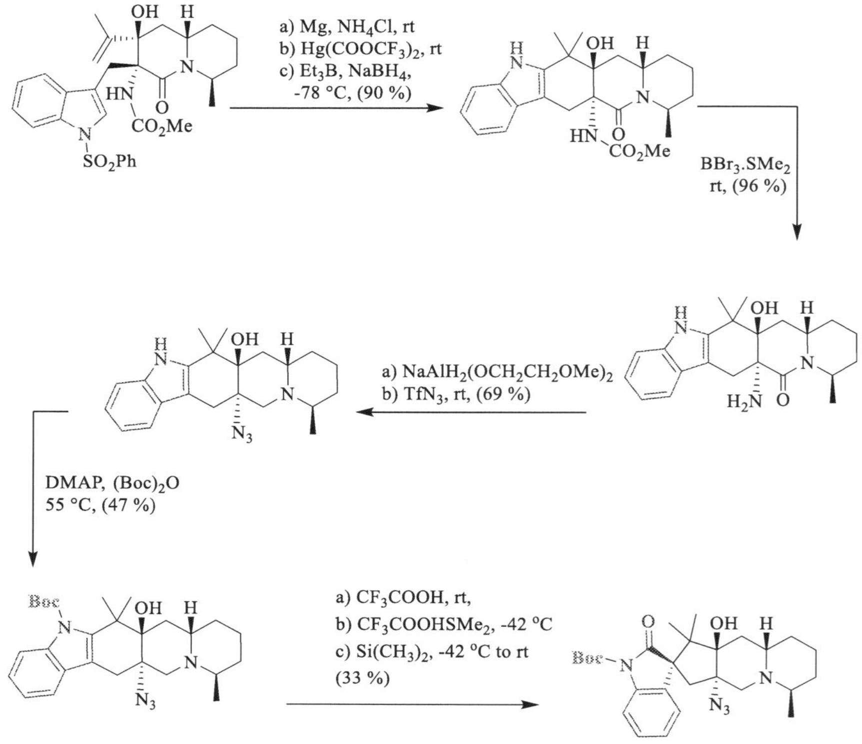

5. Citrinadin A and B

40

SCHEME 5.2 (Continued).

Key features

- Stereocontrolled synthesis of citrinadin core from scalemic, readily available starting material
- Coupling of both prepared fragments; *tert*-butyl ester and pentafluorophenyl ester give via mixed Claisen acylation to generate β-keto ester
- Diastereoselective carbonyl addition of iso-propenylmagnesium bromide to the ketocarbonyl of lactam
- Reduction of lactam followed by conversion of primary amine to an azide
- Carbamoylation of the indole nitrogen
- Oxidative rearrangement to furnish spiro core

5.1 Syntheses of citrinadins A and B spirooxindole core   **41**

5.1.3 Wood's enantioselective synthesis of (+)-citranidin A core (2016)

See Scheme 5.3.

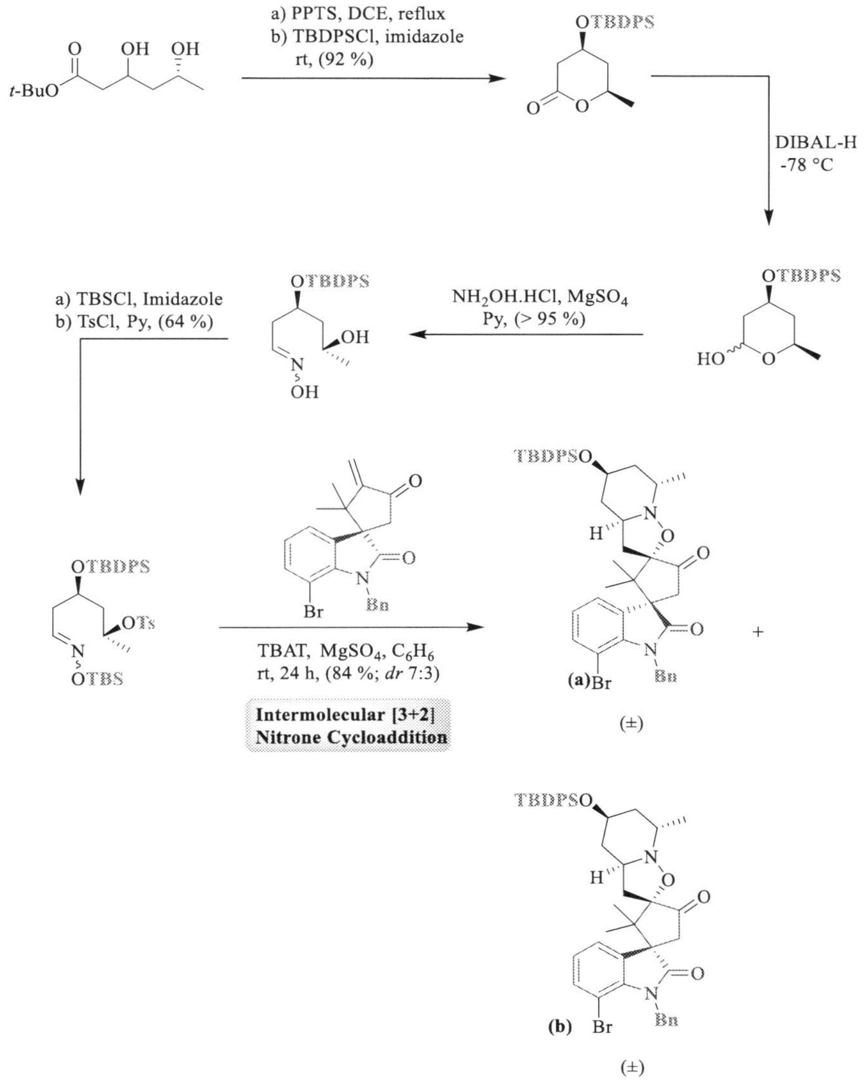

SCHEME 5.3   Wood's enantioselective synthesis of (+)-citranidin A core.[5]

Indole Alkaloids

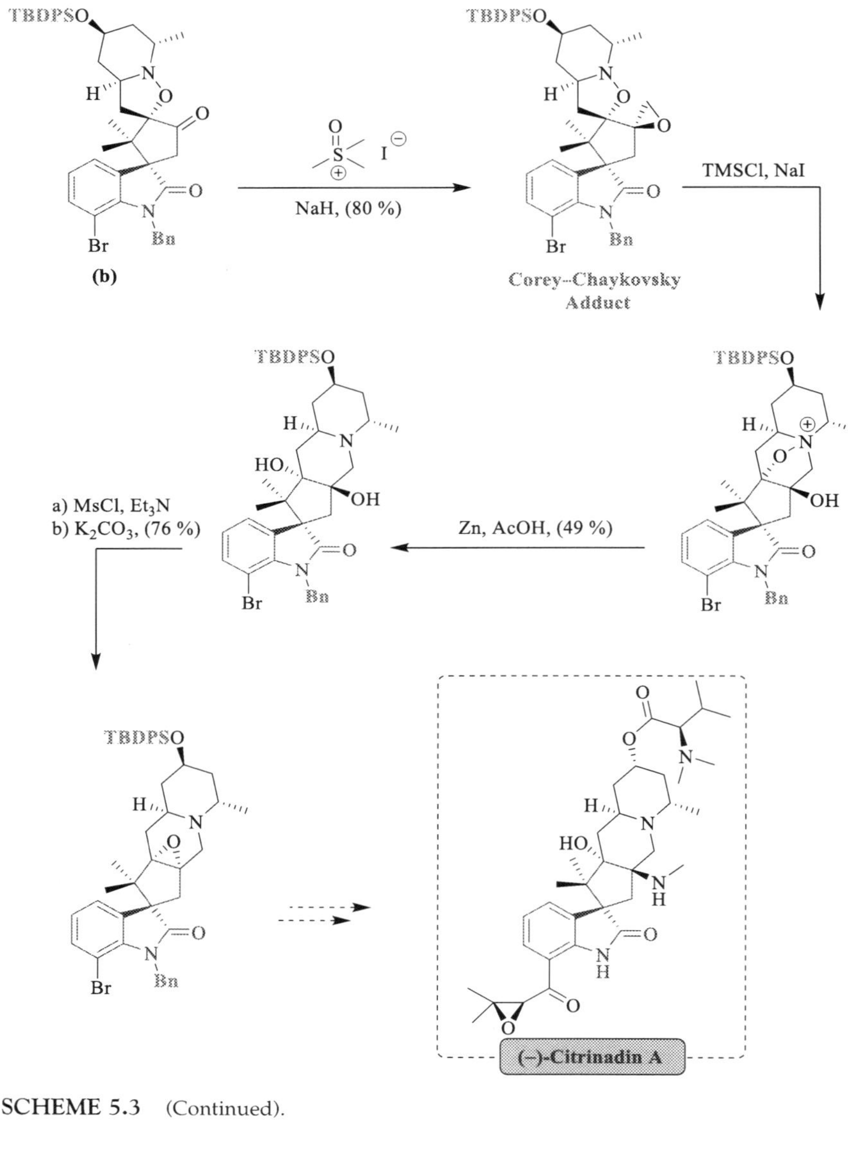

42

5. Citrinadin A and B

SCHEME 5.3 (Continued).

Key features

- Fabrication of late-stage intermediate for the synthesis of citrinadin A
- Key step involves an intermolecular [3 + 2] nitrone cycloaddition
- Synthesis of racemic nitrone precursor starting from a known diol

- Lactone synthesis from diol
- Conversion of lactone to lactol and oxime formation
- Protection of alcohol of oxime alcohol followed by activation of remaining secondary alcohol via tosylation
- Cycloaddition of racemic nitrone to already prepared racemic enone
- Conversion of cycloadded product to Corey–Chaykovsky adduct
- Fabrication of ammonium salt by intramolecular attack on the epoxide by nitrogen of isoxazolidine
- Reduction to give diol moiety
- Activation of least hindered alcohol to form desired key intermediate

5.2 Total syntheses of citrinadins A and B

5.2.1 Martin's first enantioselective total synthesis of citrinadin A (2013)

See Scheme 5.4.

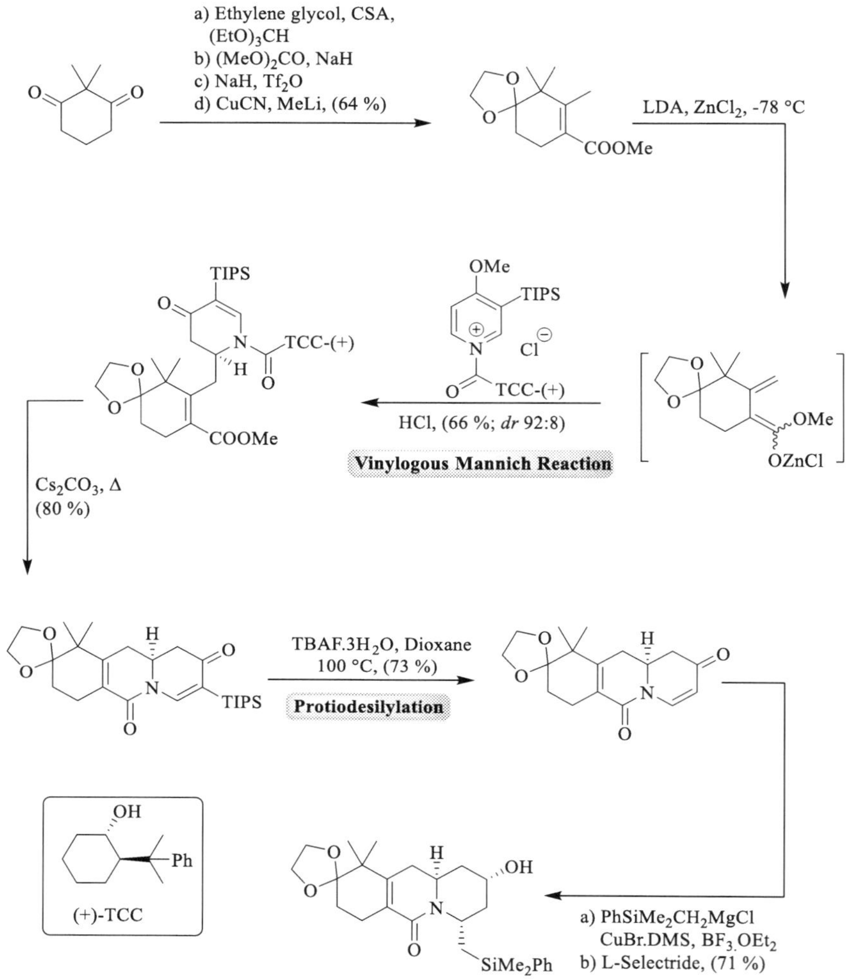

5. Citrinadin A and B

SCHEME 5.4 Martin's first enantioselective total synthesis of citrinadin A.[6]

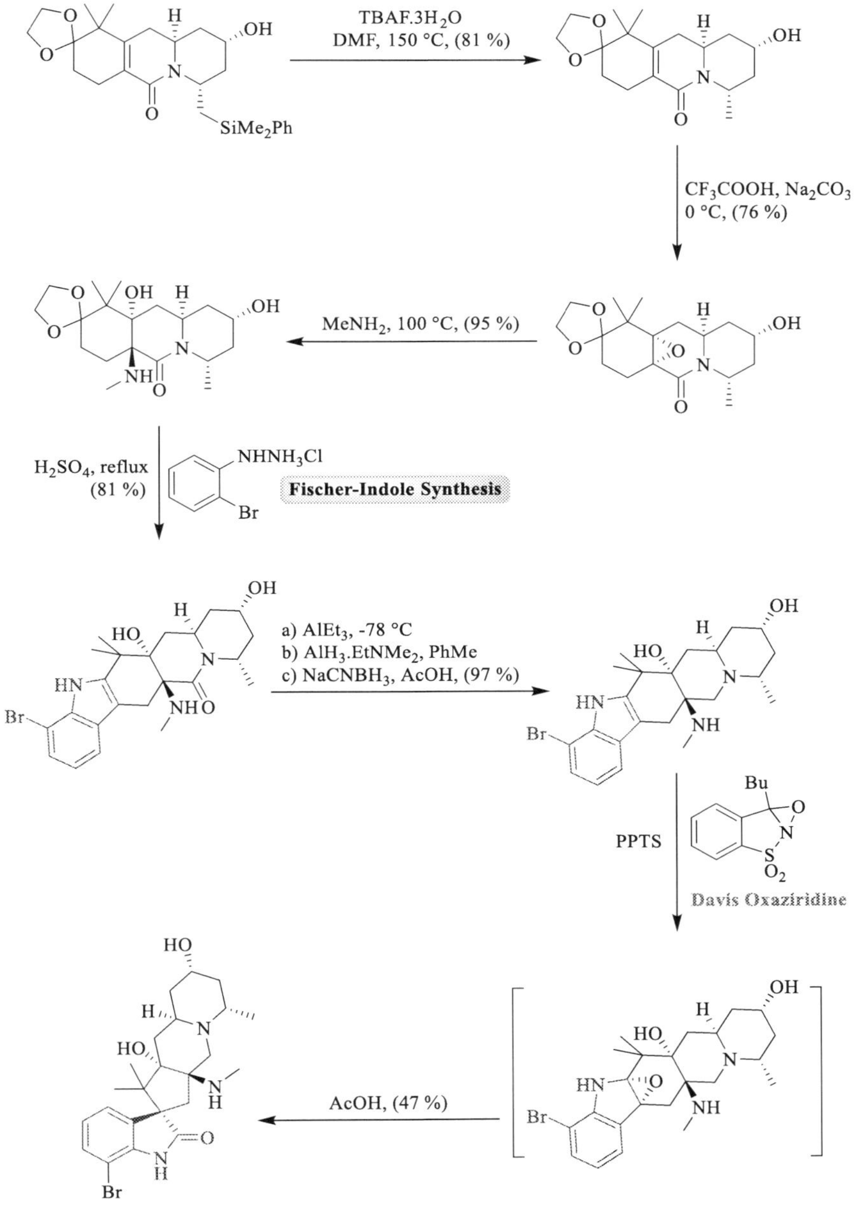

5.2 Total syntheses of citrinadins A and B

45

SCHEME 5.4 (Continued).

Indole Alkaloids

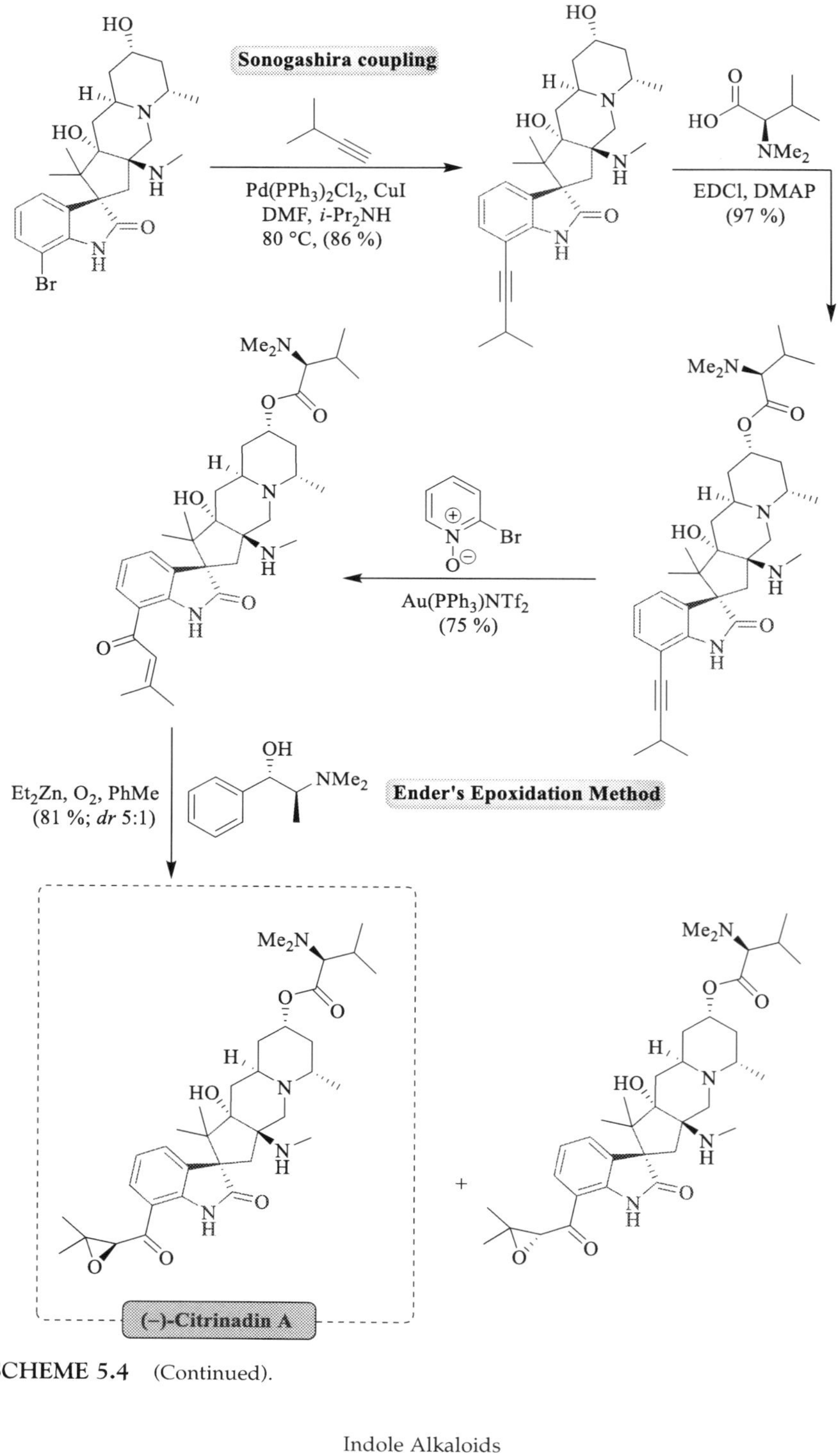

46

5. Citrinadin A and B

SCHEME 5.4 (Continued).

Indole Alkaloids

Key features

- First enantioselective total synthesis of citrinadin A from commercially available starting materials in 20 steps.
- Minimum protection/deprotection steps in the synthetic pathway.
- The synthetic sequence involves the stereoselective epoxidation/rig opening and oxidative rearrangement of an indole to construct spirooxindole.
- Further stereocenters also established in pentacyclic core of target molecule.
- Sonogashira coupling between produce spiro core intermediate and 3-methylbut-1-yne for alkyne substitution.

5.2.2 Martin's enantioselective total syntheses of (−)-citrinadin A and (+)-citrinadin B (2014)

5.2.2.1 Synthesis of (−)-citrinadin A isomer

See Scheme 5.5.

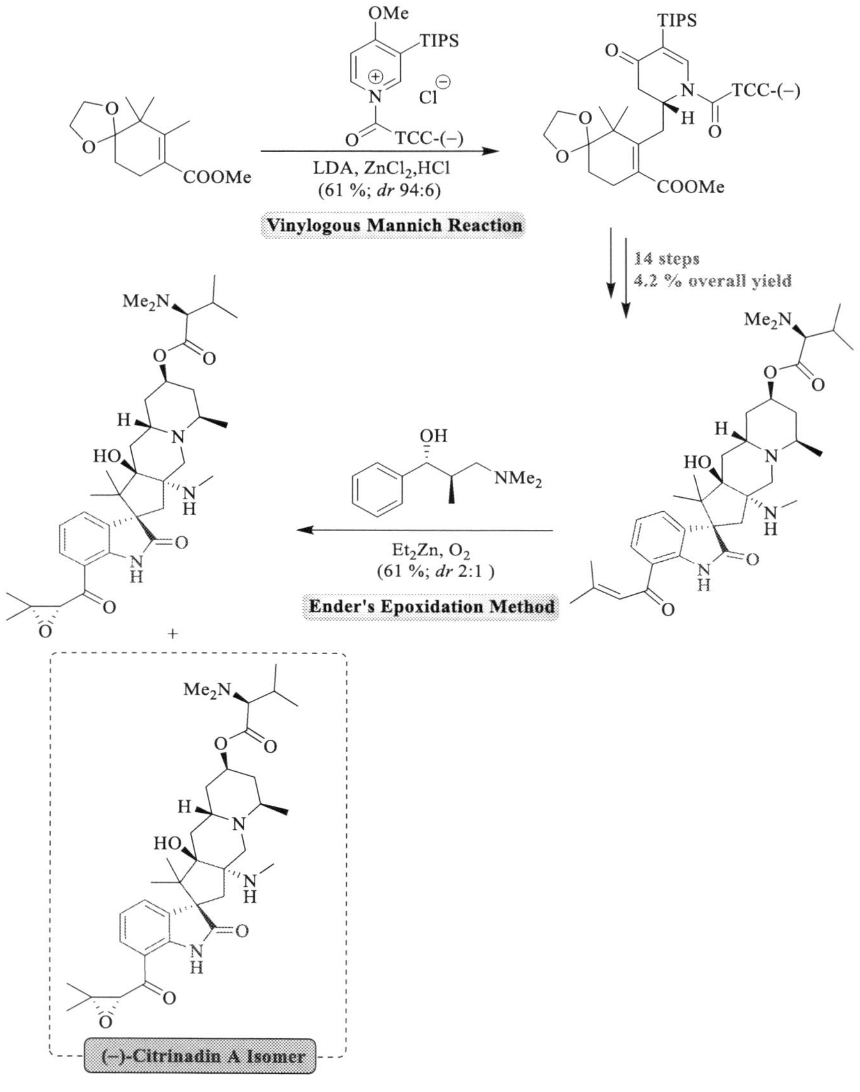

48

5. Citrinadin A and B

SCHEME 5.5 Synthesis of (−)-Citrinadin A isomer.[7]

Key features

- Conversion of commercially available 2,2-dimethylcyclohexane-1, 3-dione into 1,3-dioxolane derivative.
- Isomer was produced by using (−)-TCC substitution as chiral auxiliary in contrast to citranidin A synthesis in which (+)-TCC substitution in chiral auxiliary was used.

Indole Alkaloids

- After obtainment of tricyclic isomeric intermediate, isomer of citranidin A was obtained by same synthetic pathway as that of citranidin A.

5.2.2.2 Synthesis of (+)-citrinadin B

See Scheme 5.6.

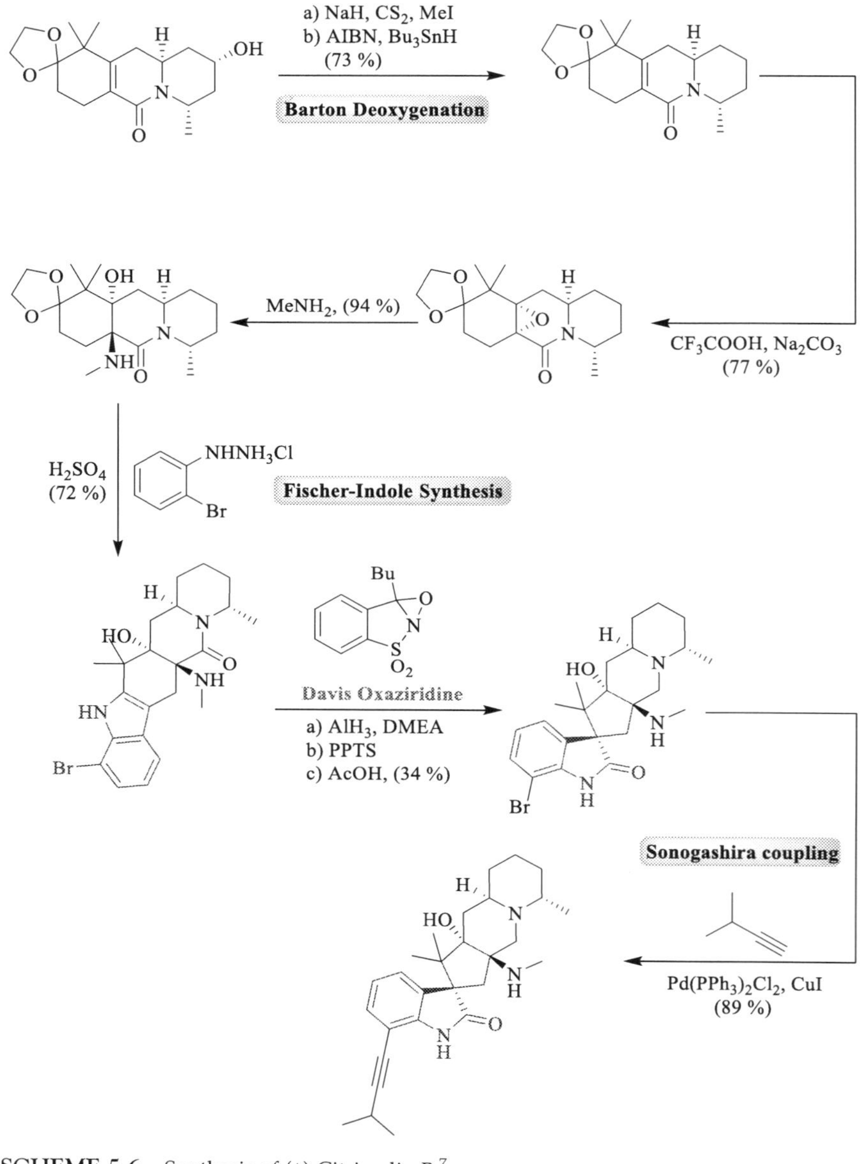

5. Citrinadin A and B

SCHEME 5.6 Synthesis of (+)-Citrinadin B.[7]

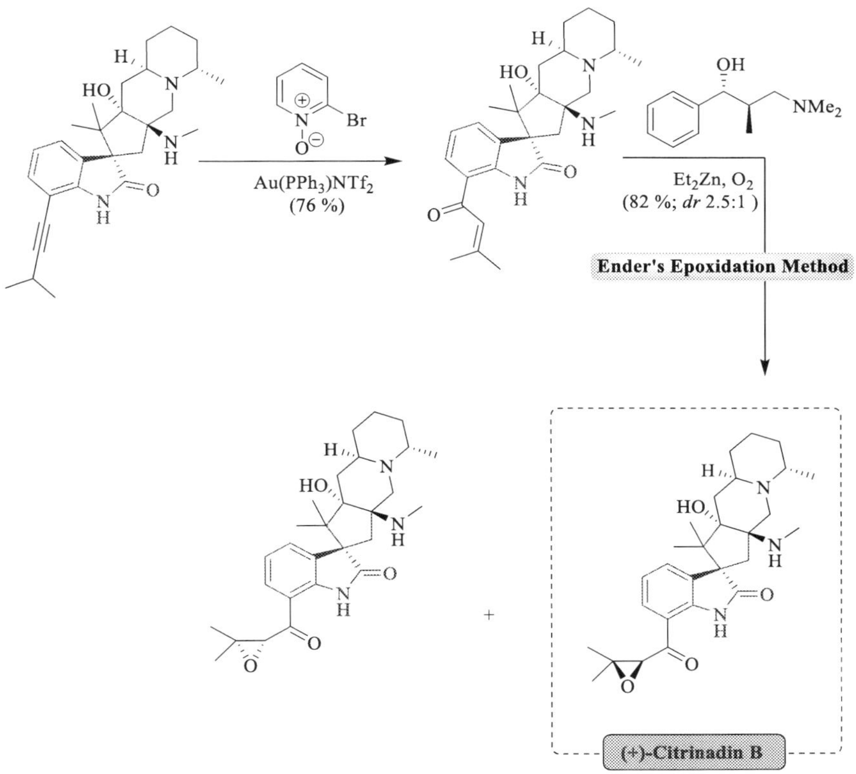

SCHEME 5.6 (Continued).

Key features

- Synthesis of (+)-citrinadin B from commercially available starting materials in 21 steps
- Production of amino alcohol derivative of lactam and Fisher indole reaction of amino alcohol with *o*-bromophenylhydrazine hydrochloride
- Spiro core construction by successive hydride reduction of the tertiary lactam, oxidation of indole with Davis' oxaziridine, and acid-catalyzed rearrangement of epoxide intermediate
- Alkyne substitution by Sonogashira coupling with 3-methylbut-1-yne
- Epoxidation of enone by Ender's process

Wood's enantioselective total synthesis of (+)-citrinadin B (2013, 2014)

See Scheme 5.7.

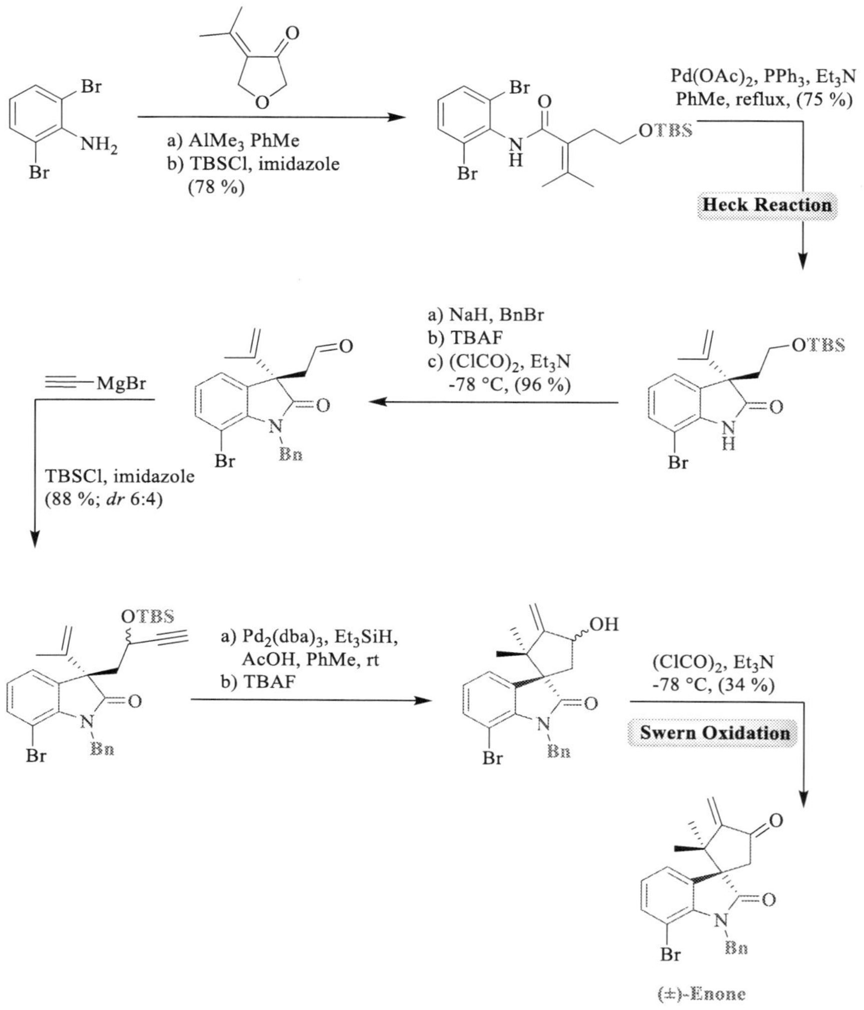

52

5. Citrinadin A and B

SCHEME 5.7 Wood's enantioselective total synthesis of (+)-Citrinadin B.[8,9]

Key features

- Treatment of dibromoaniline with lactone and trimethylaluminum
- Cyclization of resultant amide under Heck conditions to produce oxindole core
- Benzyl protection, silyl group cleavage, and alcohol oxidation
- Grignard addition

Indole Alkaloids

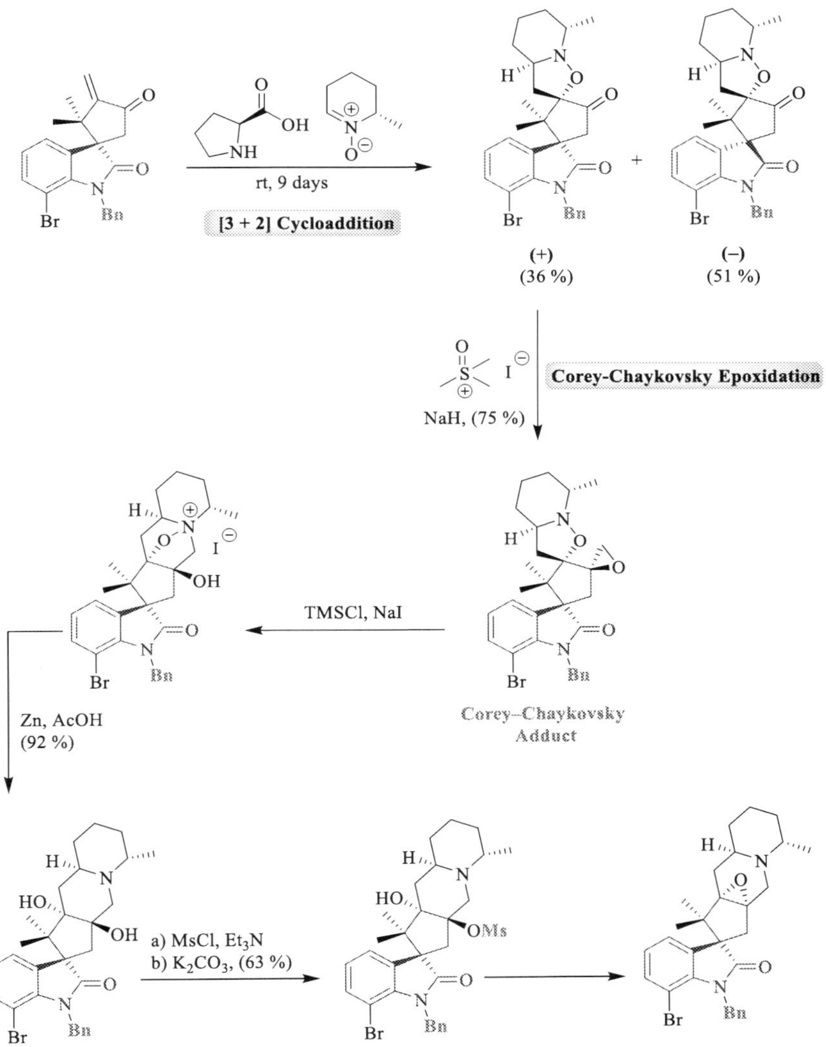

5.2 Total syntheses of citrinadins A and B

SCHEME 5.7 (Continued).

- Spirocyclization of resultant intermediate
- Swern oxidation to obtain dipolar cycloaddition substrate (±)-enone
- Intermolecular [3 + 2] cycloaddition between (±)-enone and (−)-nitrone
- Corey−Chaykovsky epoxidation
- Intramolecular epoxide opening to produce diol

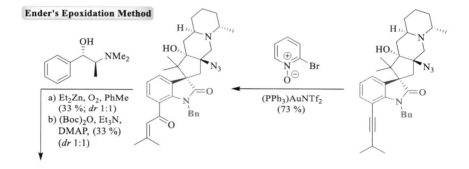

SCHEME 5.7 (Continued).

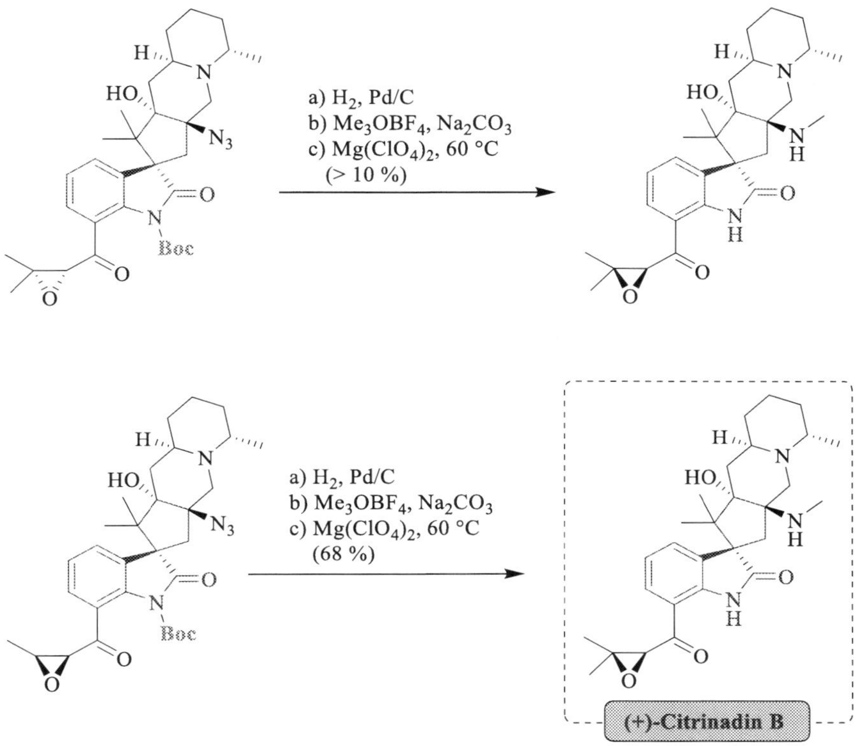

SCHEME 5.7 (Continued).

- Again epoxidation
- Alkyne coupling under Sonogashira conditions
- Gold mediated oxidation for the conversion of alkyne to an enone, thus delivering azido alcohol
- Ender's epoxidation
- Conversion of azide moiety to corresponding methylamine via azide reduction, monomethylation, and deprotection

References

1. Mugishima, T.; Tsuda, M.; Kasai, Y.; Ishiyama, H.; Fukushi, E.; Kawabata, J.; Watanabe, M.; Akao, K.; Kobayashi, J. I. Absolute Stereochemistry of Citrinadins A and B From Marine-Derived Fungus. *J. Org. Chem.* **2005,** *70* (23), 9430−9435.
2. Tsuda, M.; Kasai, Y.; Komatsu, K.; Sone, T.; Tanaka, M.; Mikami, Y.; Kobayashi, J. I. Citrinadin A, A Novel Pentacyclic Alkaloid from Marine-Derived Fungus Penicillium C Itrinum. *Org. Lett.* **2004,** *6* (18), 3087−3089.

3. Pettersson, M.; Knueppel, D.; Martin, S. F. Concise, Stereoselective Approach to the Spirooxindole Ring System of Citrinadin A. *Org. Lett.* **2007,** *9* (22), 4623−4626.
4. Guerrero, C. A.; Sorensen, E. J. Concise, Stereocontrolled Synthesis of the Citrinadin B Core Architecture. *Org. Lett.* **2011,** *13* (19), 5164−5167.
5. McCallum, M. E.; Smith, G. M.; Matsumaru, T.; Kong, K.; Enquist, J. A., Jr; Wood, J. L. Synthetic Studies Toward Citrinadin A: Construction of the Pentacyclic Core. *J. Antibiot.* **2016,** *69* (4), 331−336.
6. Bian, Z.; Marvin, C. C.; Martin, S. F. Enantioselective Total Synthesis of (−)-Citrinadin A and Revision of Its Stereochemical Structure. *J. Am. Chem. Soc.* **2013,** *135* (30), 10886−10889.
7. Bian, Z.; Marvin, C. C.; Pettersson, M.; Martin, S. F. Enantioselective Total Syntheses of Citrinadins A and B. Stereochemical Revision of Their Assigned Structures. *J. Am. Chem. Soc.* **2014,** *136* (40), 14184−14192.
8. Kong, K.; Enquist, J. A., Jr; McCallum, M. E.; Smith, G. M.; Matsumaru, T.; Menhaji-Klotz, E.; Wood, J. L. An Enantioselective Total Synthesis and Stereochemical Revision of (+)-Citrinadin B. *J. Am. Chem. Soc.* **2013,** *135* (30), 10890−10893.
9. Matsumaru, T.; McCallum, M. E.; Enquist, J. A., Jr; Smith, G. M.; Kong, K.; Wood, J. L. Synthetic Studies Toward the Citrinadins: Enantioselective Preparation of an Advanced Spirooxindole Intermediate. *Tetrahedron* **2014,** *70* (27−28), 4089−4093.

CHAPTER

6

Coerulescine and horsfiline

Horsfiline was first isolated by Bodo and coworkers in 1991 from the Malaysian tree *Horsfieldia superba* that is an important source of medicinal extracts and snuffs in local medicines.[1] Coerulescine had already been synthesized on pilot scale to synthesize horsfiline[2] and vincadifformine[3] before its isolation from natural source *Phalaris coerulescens* and characterization as spirooxindole alkaloid (Fig. 6.1) by Colegate's group in 1998.[4] The most important biological activity associated with horsfiline is that it is used as an intoxicating snuff substance.[1]

Several racemic and enantiomeric synthetic tactics have been reported for spirooxindole core construction of coerulescine and horsfiline. However, only five cases of enantiomerically enriched horsfiline are there.

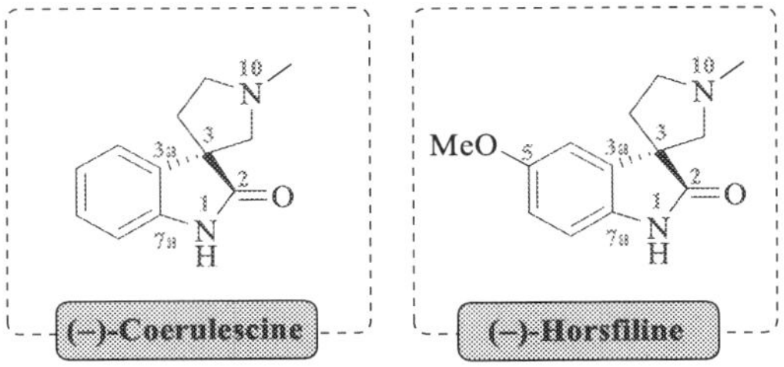

FIGURE 6.1   Structures of (−)-coerulescine and (−)-horsfiline.

Indole Alkaloids
DOI: https://doi.org/10.1016/B978-0-323-91674-5.00012-2

© 2022 Elsevier Inc. All rights reserved.

58   6. Coerulescine and horsfiline

6.1 Total synthesis of coerulescine and horsfiline

6.1.1 Jones' total synthesis of horsfiline via radical cyclization (1992)

See Scheme 6.1.

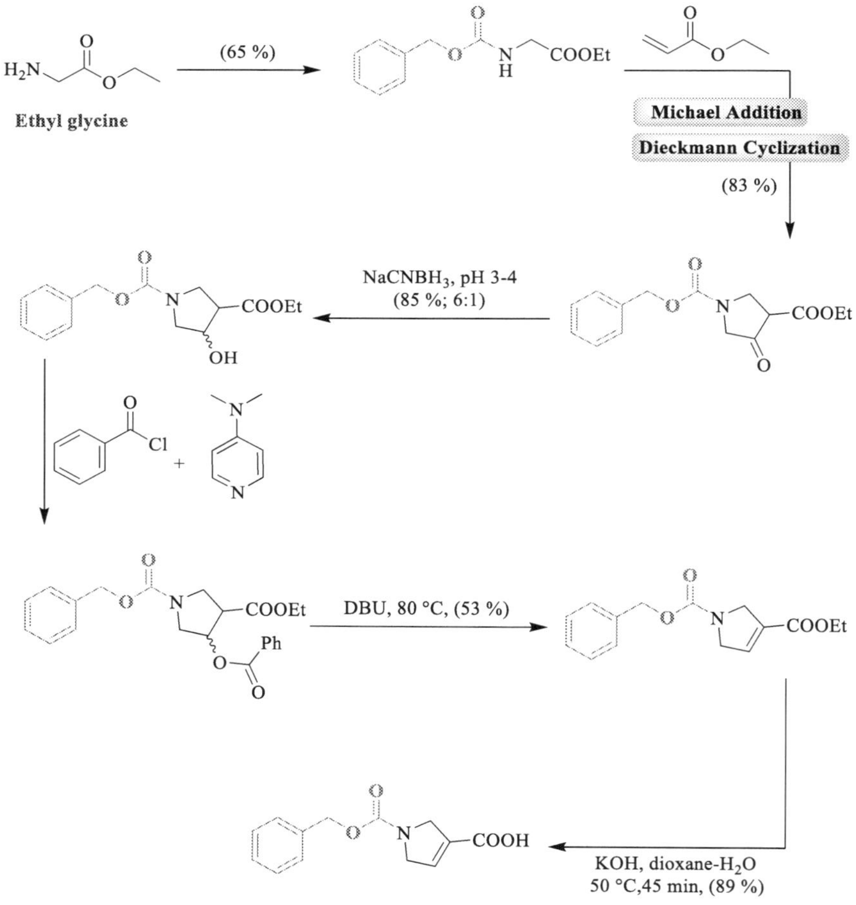

SCHEME 6.1   Jones' total synthesis of horsfiline via radical cyclization.[5]

Indole Alkaloids

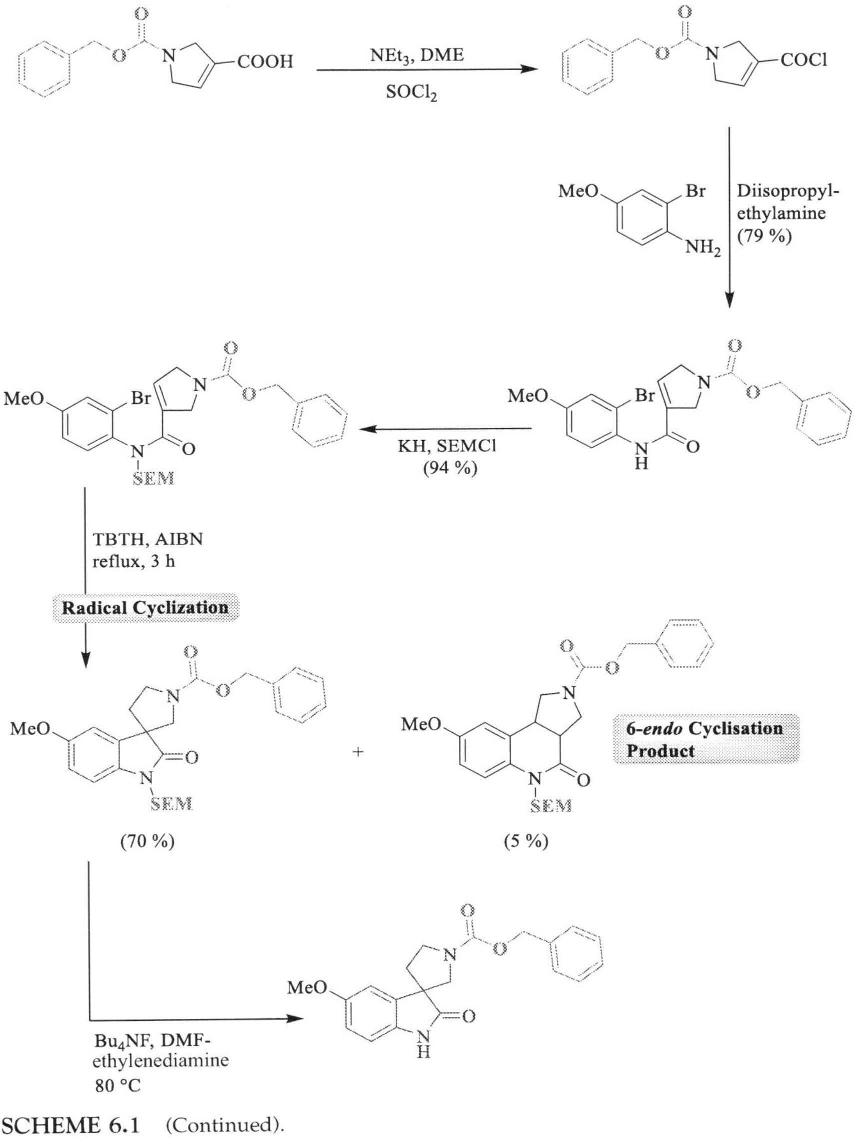

6.1 Total synthesis of coerulescine and horsfiline

SCHEME 6.1 (Continued).

Key features

- Protection of amino acid ethyl glycine
- Michael addition followed by Dieckmann cyclization of protected ethyl glycine with ethyl acrylate

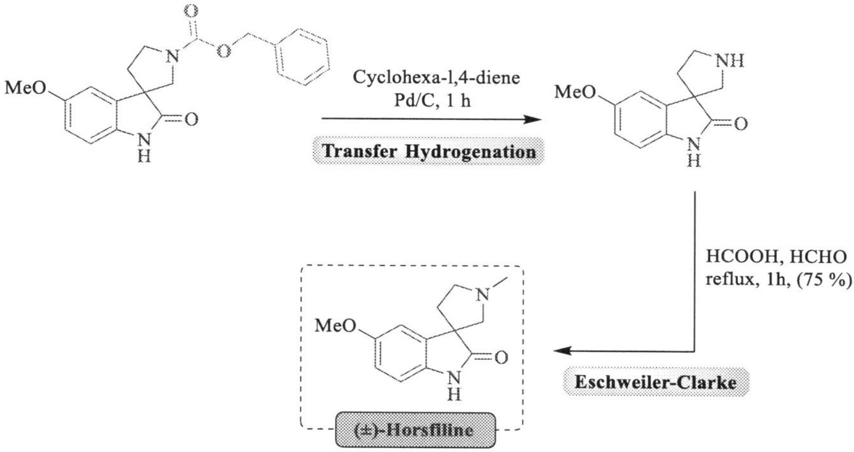

60

6. Coerulescine and horsfiline

SCHEME 6.1 (Continued).

- Reduction of pyrrolidinone diastereomeric intermediate
- Formation of α,β-unsaturated ester by hydroxyl elimination
- Ester hydrolysis
- Formation of acid chloride
- Condensation of acid chloride with 2-bromo-4-methoxyaniline to form α,β-unsaturated amide
- Amide's nitrogen protection, treatment of protected amide with radical initiator to form spiro core via 5-*exo* cyclization followed by amide deprotection
- Transfer hydrogenation to remove benzyl formate protection done on amino acid initially
- Eschweiler—Clarke conditions for *N*-methylation and to achieve horsfiline final product

Indole Alkaloids

6.1.2 Fuji's asymmetric total synthesis of (−)-horsfiline (1999)

See Scheme 6.2.

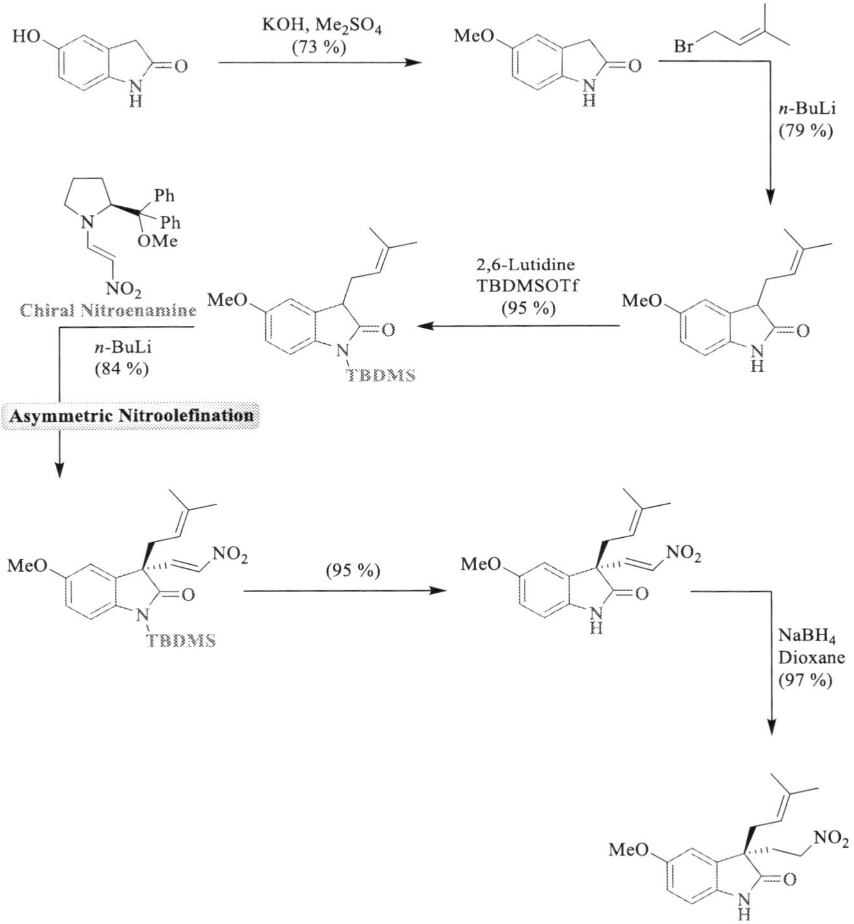

SCHEME 6.2   Fuji's asymmetric total synthesis of (−)-horsfiline.[6]

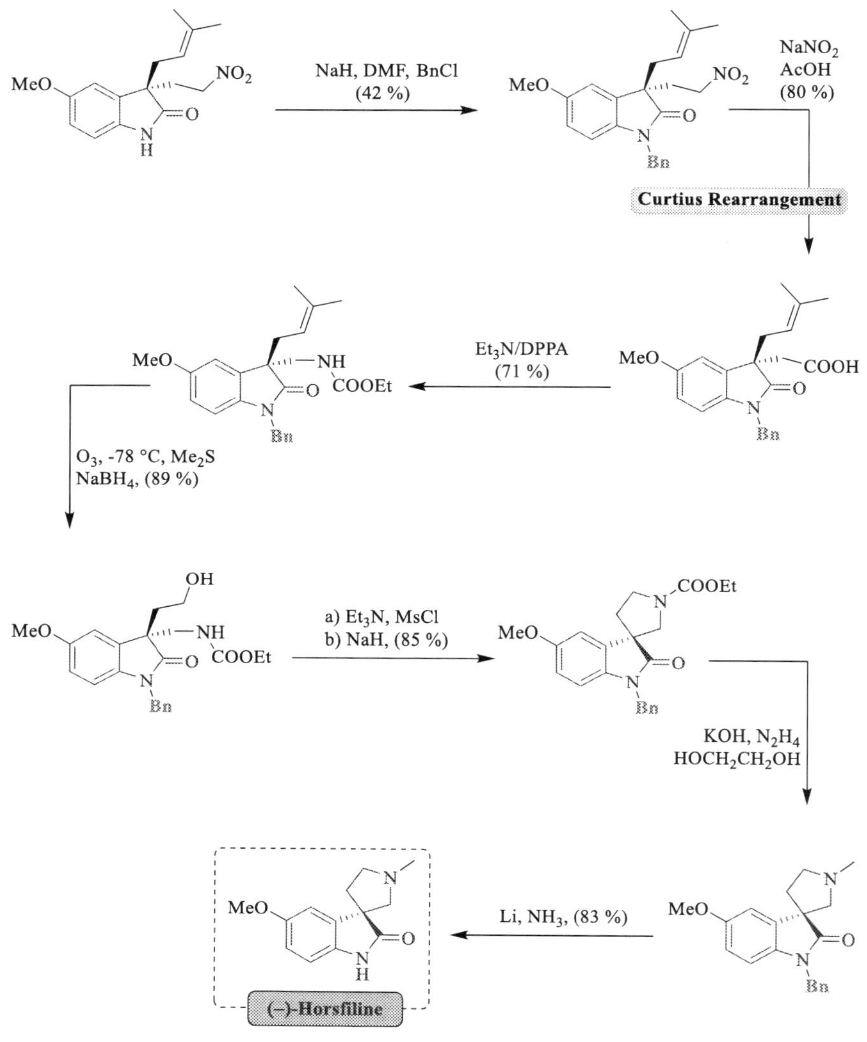

6. Coerulescine and horsfiline

SCHEME 6.2 (Continued).

Key features

- Selective *O*-methylation of 5-hydroxy-2-oxindole
- Asymmetric nitroolefination using chiral nitroenamine
- Reduction of nitroolefin
- Conversion of nitroalkane to carboxylic acid
- Thermal Curtius rearrangement to furnish corresponding ethoxy carbamate

6.1 Total synthesis of coerulescine and horsfiline

- Ozonolysis and subsequent reduction of resultant aldehyde
- Spirocyclization

6.1.3 Carreira's racemic total synthesis of (±)-horsfiline (2000)

See Scheme 6.3.

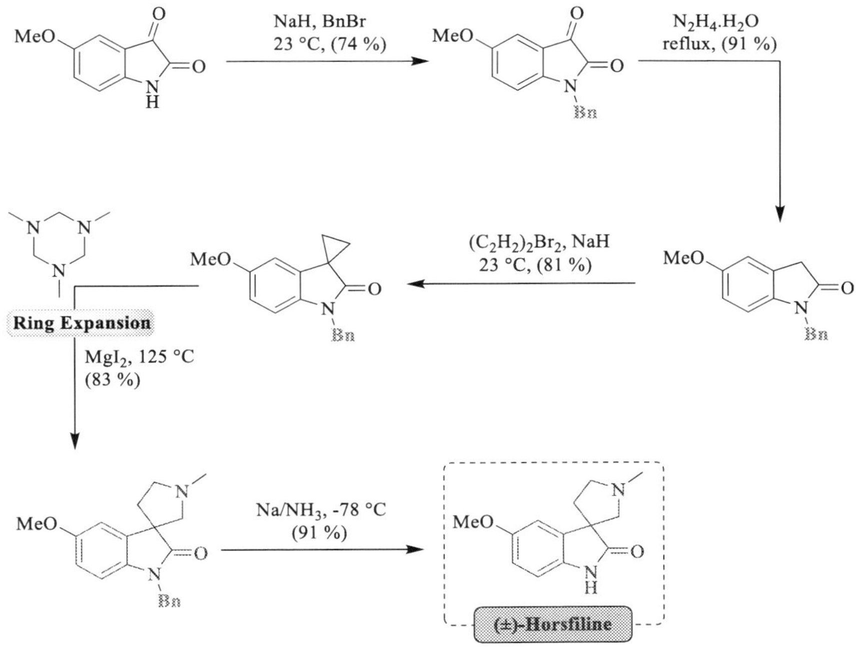

SCHEME 6.3   Carreira's racemic total synthesis of (±)-horsfiline.[7]

Key features

- Five steps synthesis of racemic horsfiline from commercially available isatin derivative in 41% overall yield
- *N*-Benzylation of substituted isatin
- Keto group reduction
- Spiro cyclopropanation
- Pyrrolidine ring at spiro fusion is obtained by cyclopropane ring expansion with magnesium iodide

6.1.4 Masanori's total synthesis of (±)-coerulescine (2000)

See Scheme 6.4.

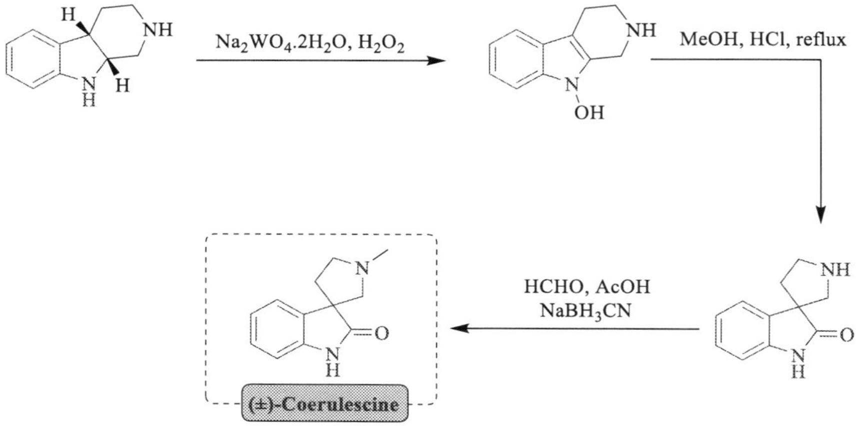

SCHEME 6.4 Masanori's total synthesis of (±)-coerulescine.[8]

Key features

- Oxidation of 1,2,3,4,4a,9a-hexahydro-β-carbolines
- Rearrangement reaction of 1,2,3,4-tetrahydro-9-hydroxy-β-carbolines to generate spiro connection
- *N*-Methylation

6.1.5 Palmisano's asymmetric total synthesis of (−)-horsfiline (2001)

See Scheme 6.5.

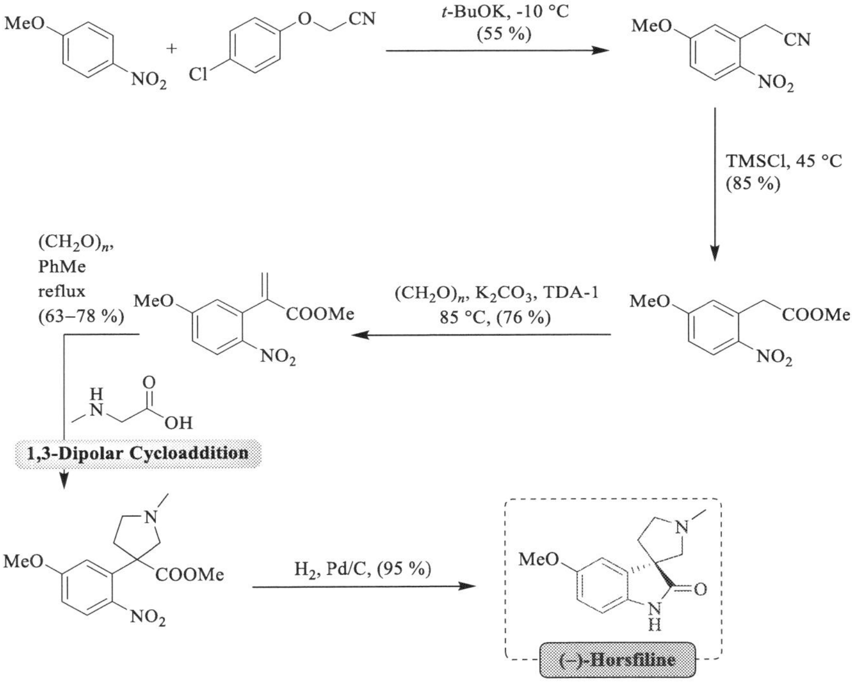

SCHEME 6.5 Palmisano's asymmetric total synthesis of (−)-horsfiline.[9]

Key features

- Intermolecular [3 + 2] annulation of azomethine ylide with 2(2-nitrophenyl)acrylate dienophile
- Reductive heterocyclization to afford spirooxindole ring system

6.1.6 Selvakumar's racemic total synthesis of (±)-coerulescine and (±)-horsfiline (2002)

See Scheme 6.6.

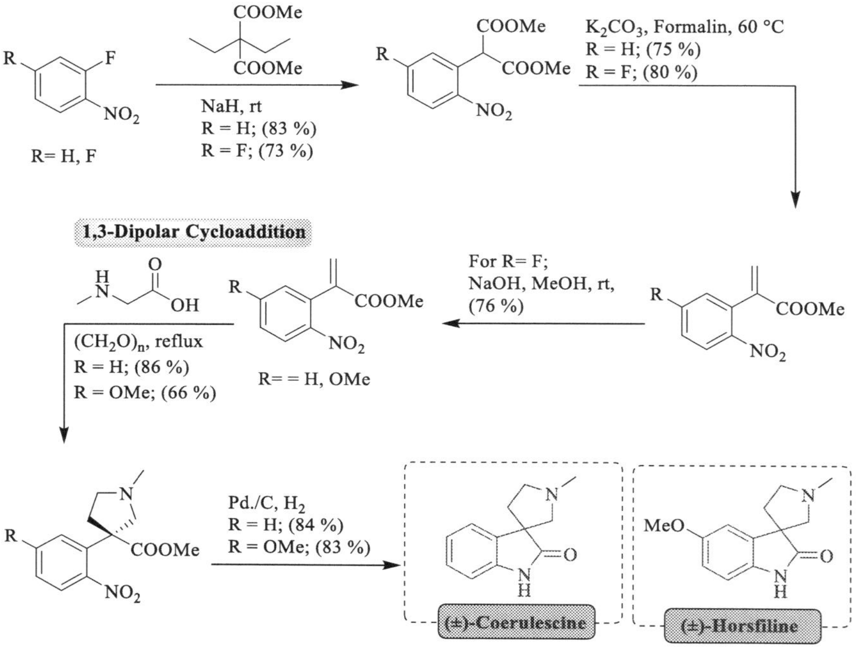

SCHEME 6.6 Selvakumar's racemic total synthesis of (±)-coerulescine and (±)-horsfiline.[10]

Key features

- Short route synthesis of (±)-coerulescine and (±)-horsfiline from commercially available aromatic nitro compounds.
- Overall yield of (±)-coerulescine is 45% and that of (±)-horsfiline is 24%.
- Aromatic nucleophilic substitution of dimethyl malonate onto substituted nitrobenzene.
- Treatment of nitromalonate with aqueous formaldehyde to yield nitro aryl propenoate via decarboxylation and respective aldol condensation mechanism.
- 1,3-Dipolar cycloaddition of nitro-ester to N-methylazomethine ylide to afford pyrrolidine nitro-ester.
- Hydrogenation leading to cyclization.

6.1.7 Murphy's racemic total synthesis of (±)-coerulescine and (±)-horsfiline (2003)

See Scheme 6.7.

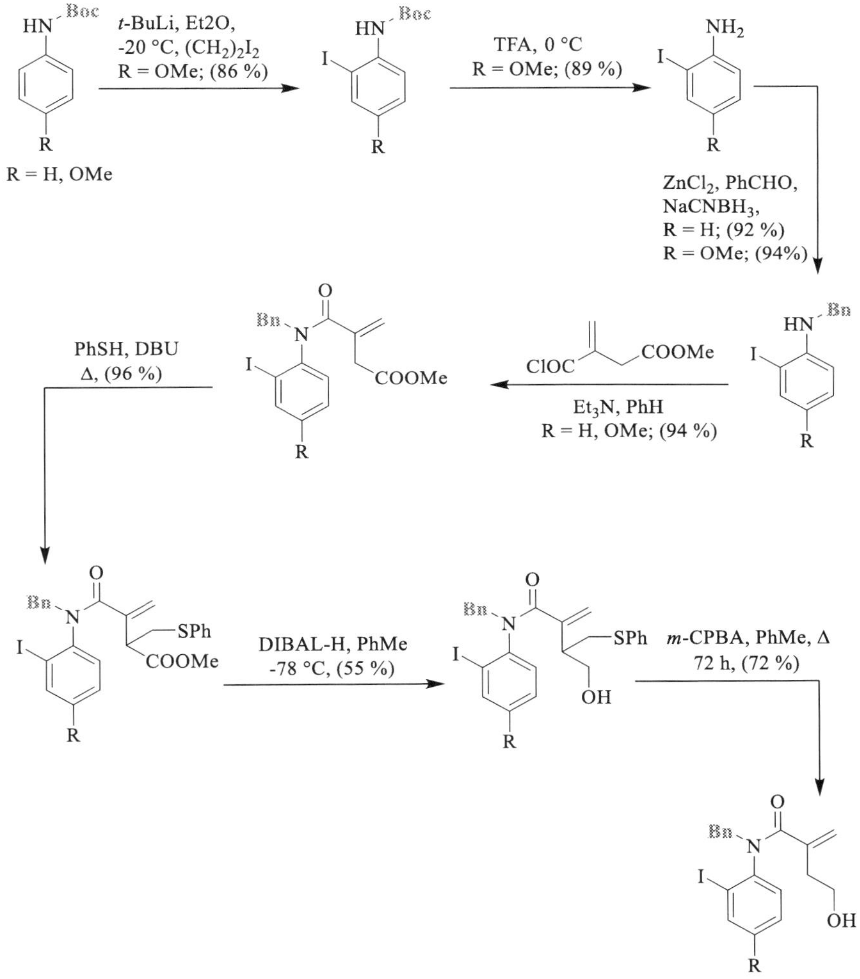

SCHEME 6.7   Murphy's racemic total synthesis of (±)-coerulescine and (±)-horsfiline.[11]

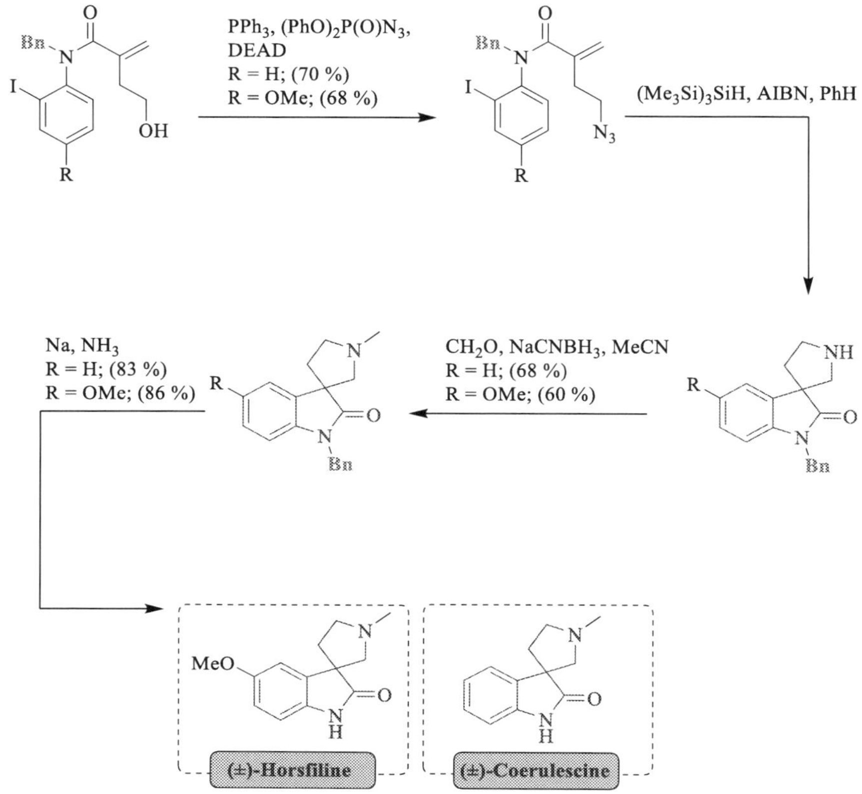

SCHEME 6.7 (Continued).

Key features

- Key step is tandem radical cyclization of iodoaryl alkenyl azides.
- *p*-Anisidine derivatives were iodinated with 1,2-diiodoethane.
- Reductive alkylation with benzaldehyde.
- Conversion of primary amine to secondary amine.
- Condensation of benzylamine with acid chloride.
- Reduction of ester group in amide to alcohol.
- Conversion to azide and cyclization.

6.1 Total synthesis of coerulescine and horsfiline

6.1.8 Chang's racemic total synthesis of (±)-coerulescine (2005)

See Scheme 6.8.

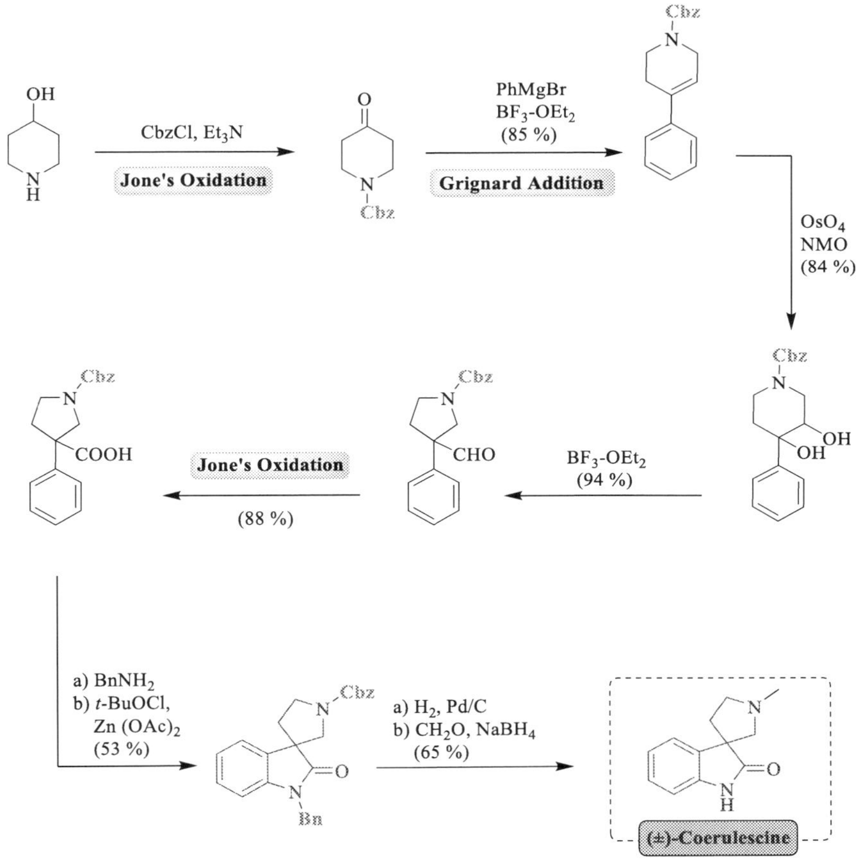

SCHEME 6.8 Chang's racemic total synthesis of (±)-coerulescine.[12]

Key features

- Synthesis of 3,4-dihydroxy-4-phenylpiperidine from commercially available 4-hydroxypiperidine via benzyloxycarbonylation, Jone's oxidation, Grignard addition, dehydration of resulting tertiary alcohols, dihydroxylation of olefins, and rearrangement of diols, respectively

Indole Alkaloids

70

6. Coerulescine and horsfiline

- Lewis acid-catalyzed rearrangement of 3,4-dihydroxy-4-phenylpiperidine to form pyrrolidine ring
- Intramolecular electrophilic cyclization of pyrrolidine to form spirocyclic oxindole ring skeleton

6.1.9 Trost's asymmetric total synthesis of (−)-horsfiline (2006)

See Scheme 6.9.

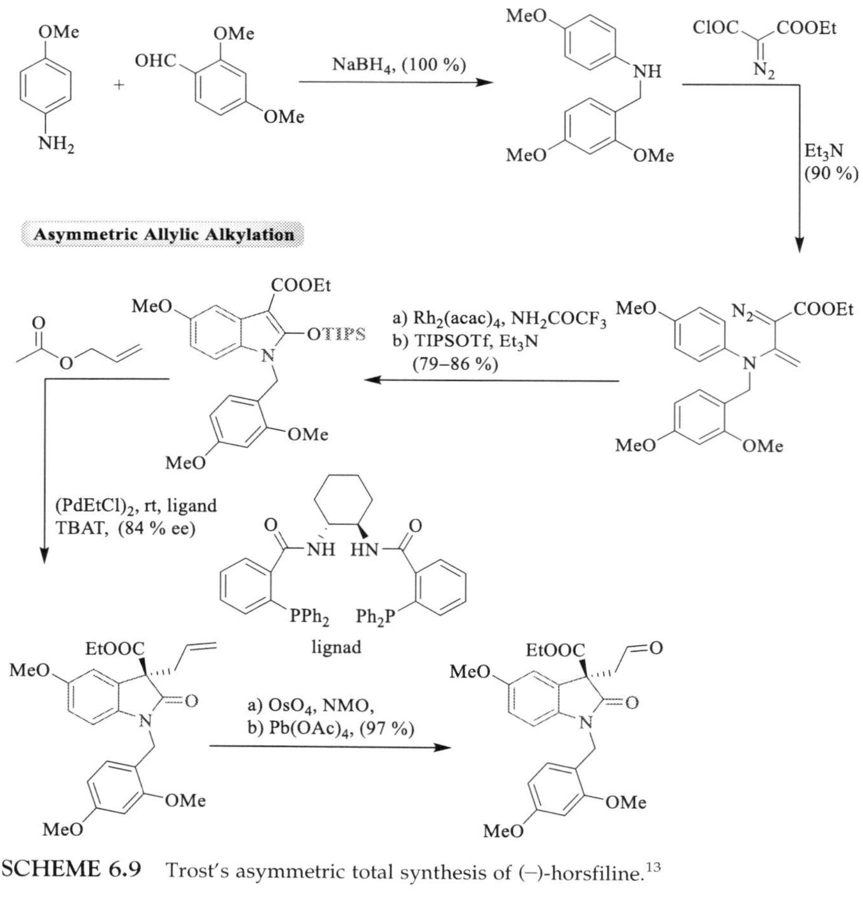

SCHEME 6.9 Trost's asymmetric total synthesis of (−)-horsfiline.[13]

Indole Alkaloids

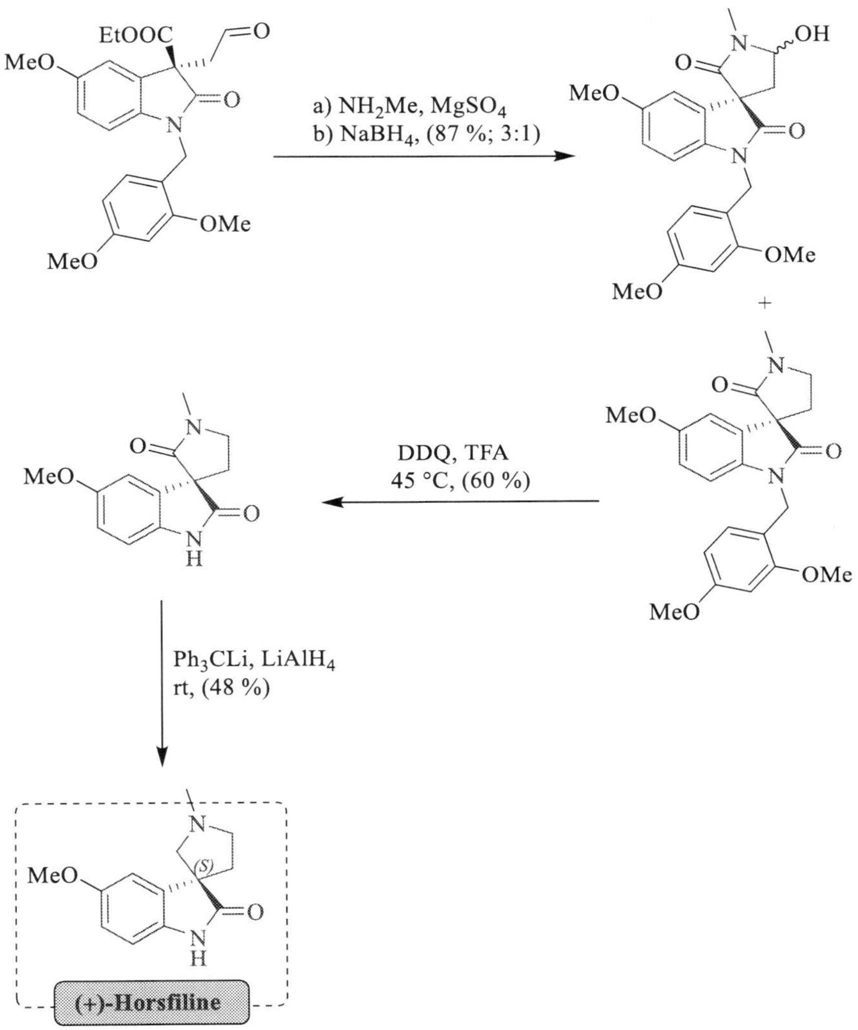

6.1 Total synthesis of coerulescine and horsfiline

SCHEME 6.9 (Continued).

Key features

- Palladium-catalyzed asymmetric allylic alkylation on an ester substituent at the third position of indole nucleus
- Oxidative cleavage of allyl group followed by reductive amination to introduce nitrogen of pyrrolidine
- Cyclization via reductive amination or S_N2 substitution to construct spiro core

6.1.10 Neuville and Zhu's total synthesis of horsfiline (2009)

See Scheme 6.10.

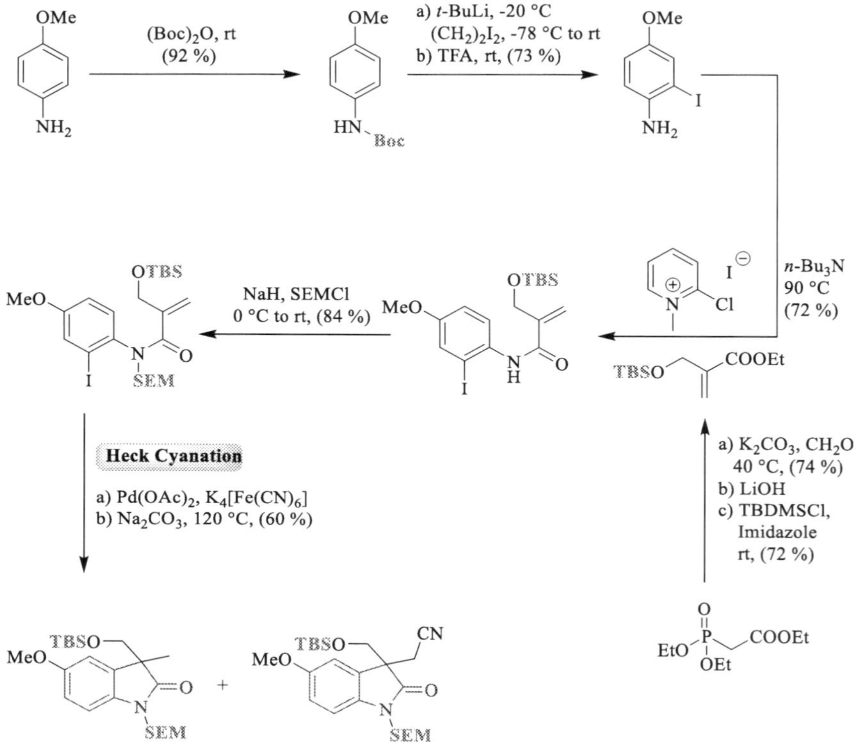

SCHEME 6.10 Neuville and Zhu's total synthesis of horsfiline.[14]

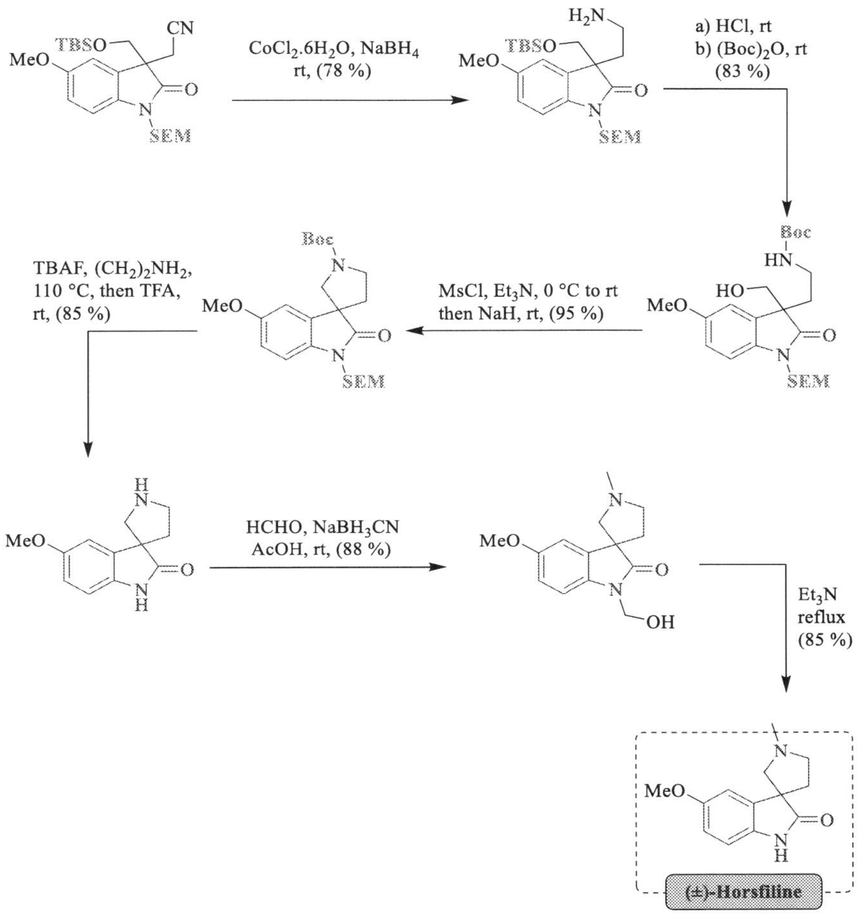

6.1 Total synthesis of coerulescine and horsfiline

73

SCHEME 6.10 (Continued).

Key features

- Preparation of protected α-hydroxymethylacrylate
- Iodide substitution to *p*-anisidine derivative
- Coupling of both prepared compounds by Mukaiyama's reagent
- Palladium-catalyzed domino intramolecular enantioselective Heck cyanation of resultant coupled compound
- Construction of the pyrrolidine ring to form spiro core

6.1.11 Maison's total synthesis of (±)-horsfiline (2010)

See Scheme 6.11.

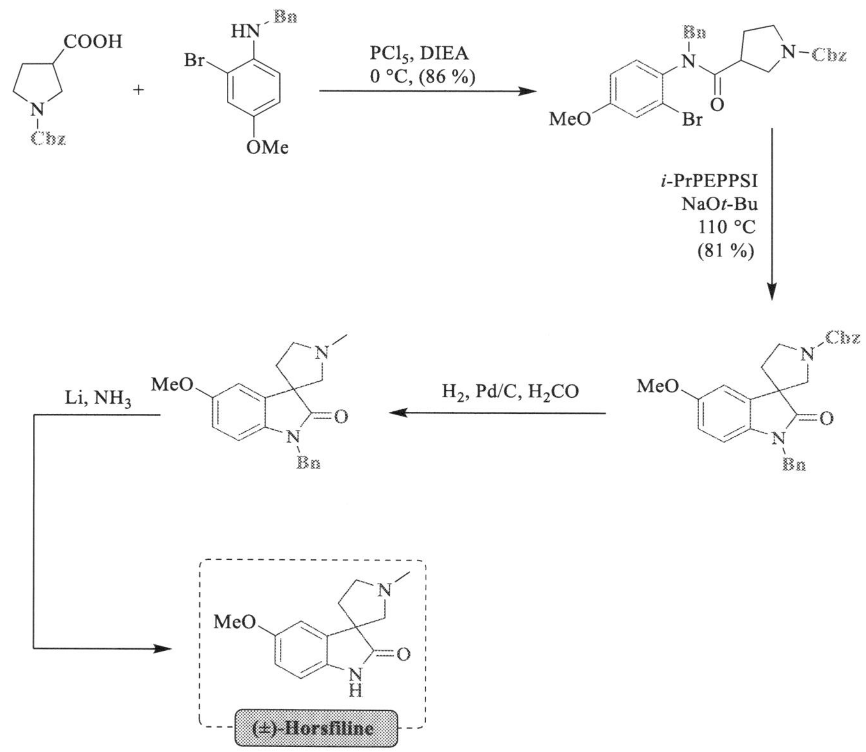

SCHEME 6.11 Maison's total synthesis of (±)-horsfiline.[15]

Key features

- Short synthesis of horsfiline in only four steps starting from commercially available amino acid
- Coupling of commercially available Cbz-protected pyrrolidine-2-carboxylic acid with an anilide
- Palladium-catalyzed α-arylation of resultant compound to form spirooxindole core structure

6.1.12 Kulkarni's total synthesis of (±)-coerulescine and (±)-horsfiline (2010)

See Scheme 6.12.

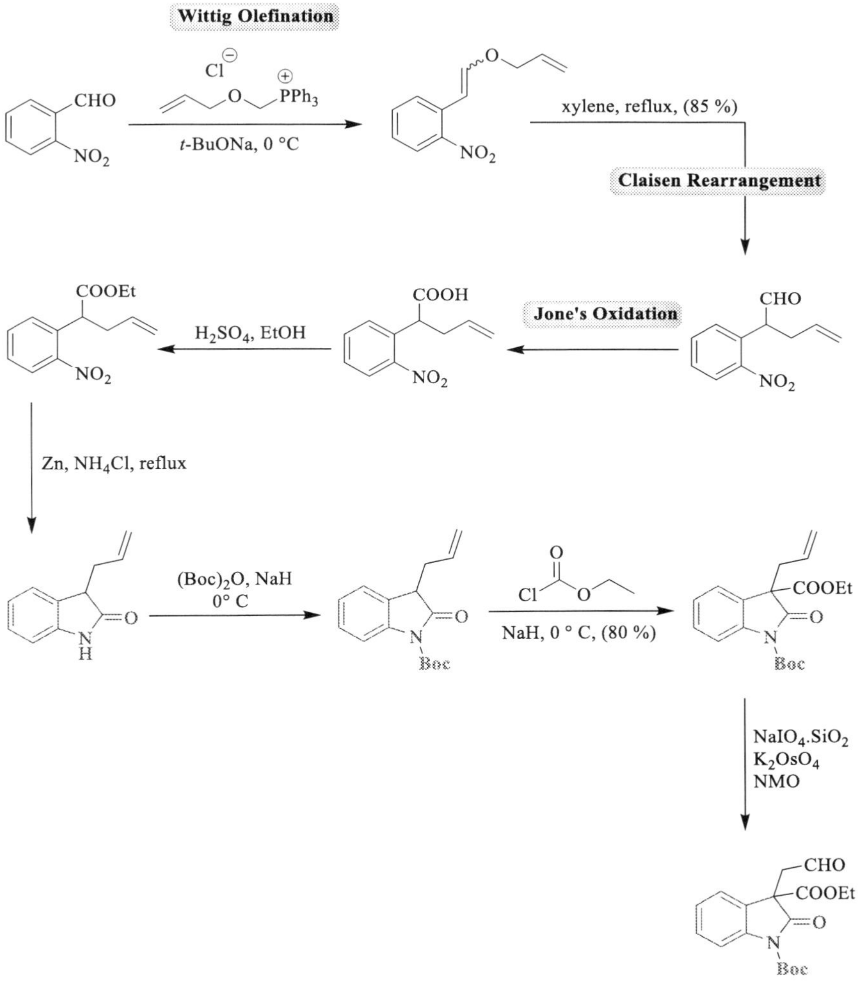

SCHEME 6.12 Kulkarni's total synthesis of (±)-coerulescine and (±)-horsfiline.[16]

Indole Alkaloids

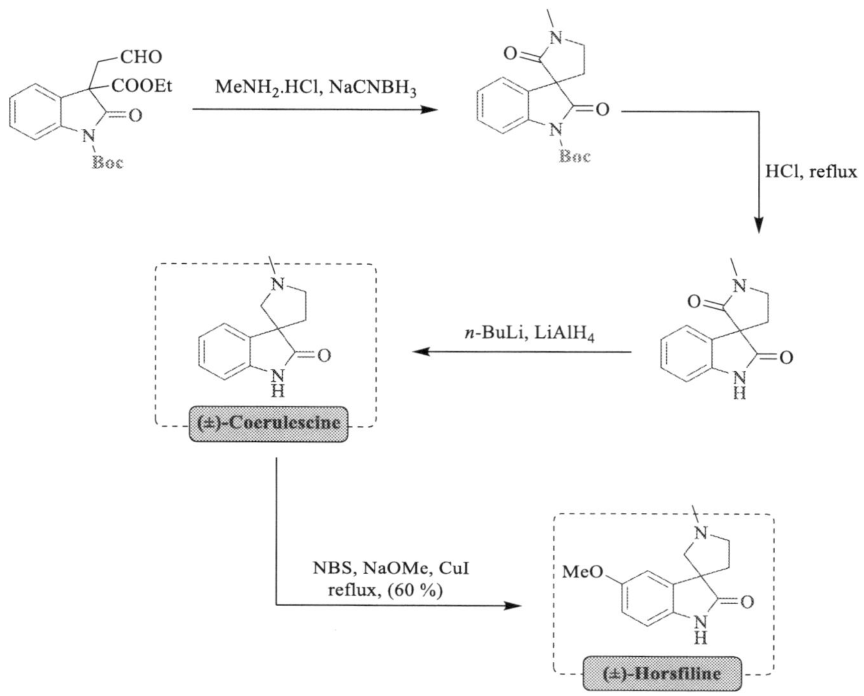

76

6. Coerulescine and horsfiline

SCHEME 6.12 (Continued).

Key features

- Wittig olefination of *o*-nitrobenzaldehyde
- Claisen rearrangement
- Jone's oxidation and esterification
- Cyclization to furnish oxindole
- *N*-Protection and acylation
- Oxidative cleavage of allyl group
- Reductive amination
- Chemoselective reduction of amide to yield (±)-coerulescine
- Conversion of (±)-coerulescine to (±)-horsfiline

6.1.13 White's racemic total synthesis of (±)-coerulescine and (±)-horsfiline (2010)

See Scheme 6.13.

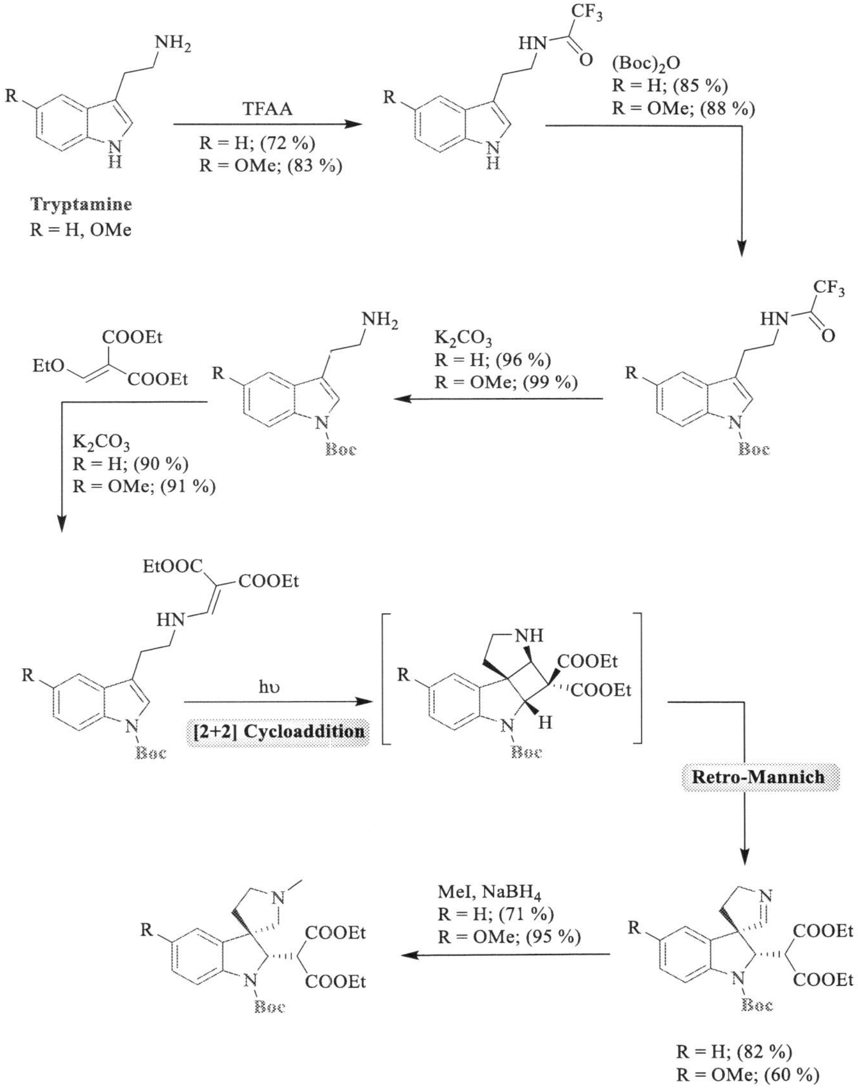

SCHEME 6.13 White's racemic total synthesis of (±)-coerulescine and (±)-horsfiline.[17]

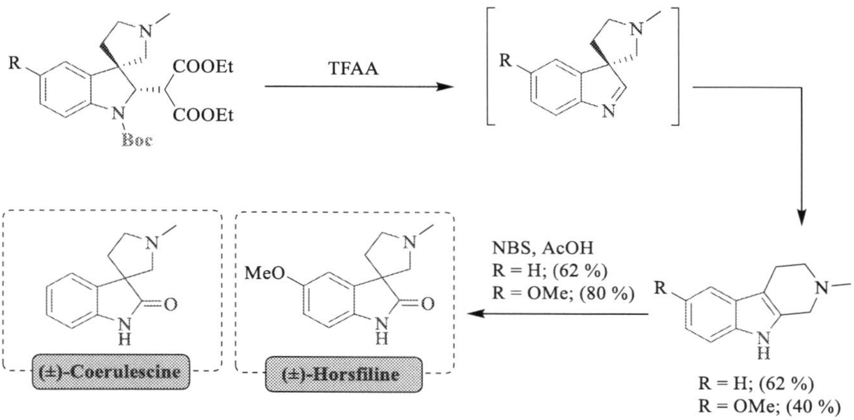

6. Coerulescine and horsfiline

SCHEME 6.13 (Continued).

Key features

- Condensation of protected 6-methoxytryptamine with β-ethoxymethylidenemalonate to yield photo substrate
- Intramolecular [2 + 2] photocycloaddition and in situ retro-Mannich fragmentation to yield spiropyrrolidine
- Second spontaneous retro-Mannich fragmentation
- Oxidative rearrangement to furnish final structure

Indole Alkaloids

6.1 Total synthesis of coerulescine and horsfiline

79

6.1.14 Kim's total synthesis of (−)-coerulescine (2012)

See Scheme 6.14.

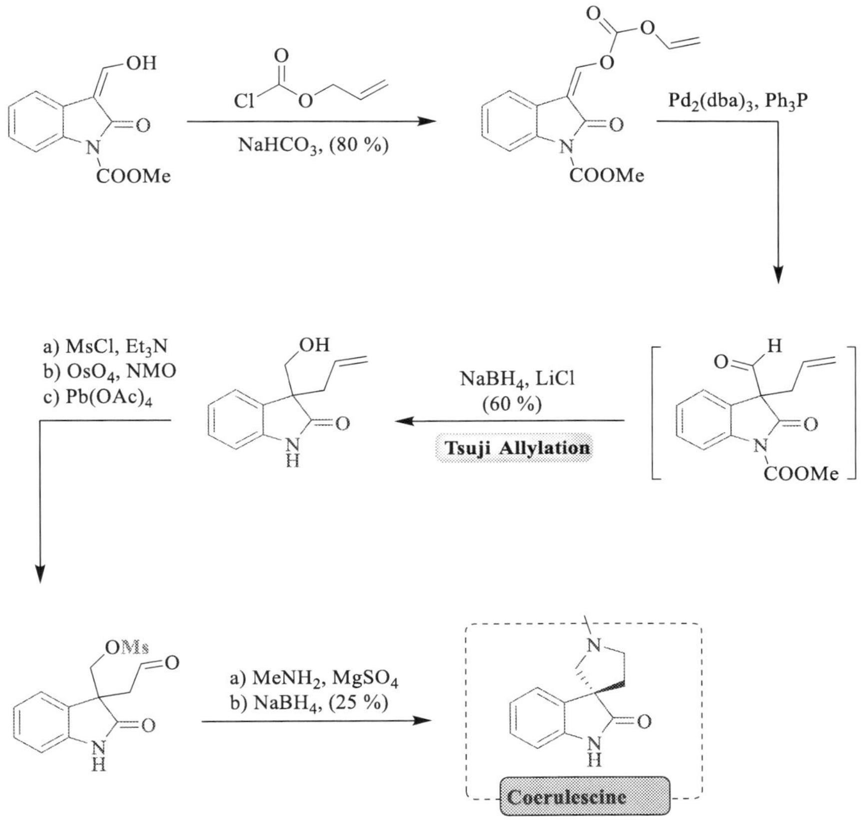

SCHEME 6.14 Kim's total synthesis of (−)-coerulescine.[18]

Key features

- Synthesis of allyl enol carbonate precursor via an established reaction
- Tsuji allylation to afford allylic aldehyde
- Deformylation
- Reductive amination and spirocyclization

Indole Alkaloids

80

6.1.15 Comesse's racemic total synthesis of (±)-coerulescine (2013)

See Scheme 6.15.

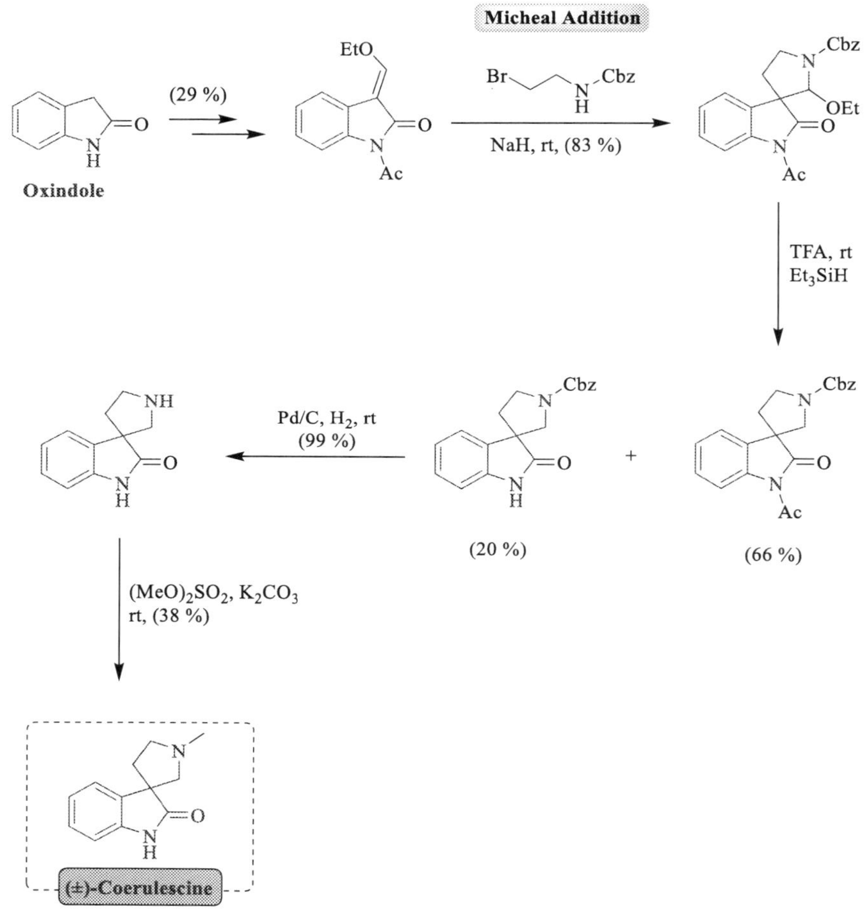

SCHEME 6.15 Comesse's racemic total synthesis of (±)-coerulescine.[19]

Key features

- Total synthesis of (±)-coerulescine achieved in seven steps and 8% overall yield starting from commercially available oxindole
- Key step includes tandem aza-Michael initiated ring closure process on ethoxymethylene oxindole obtained from commercially available oxindole in two steps

6.1.16 Park's enantioselective total synthesis of (−)-horsfiline (2013)

See Scheme 6.16.

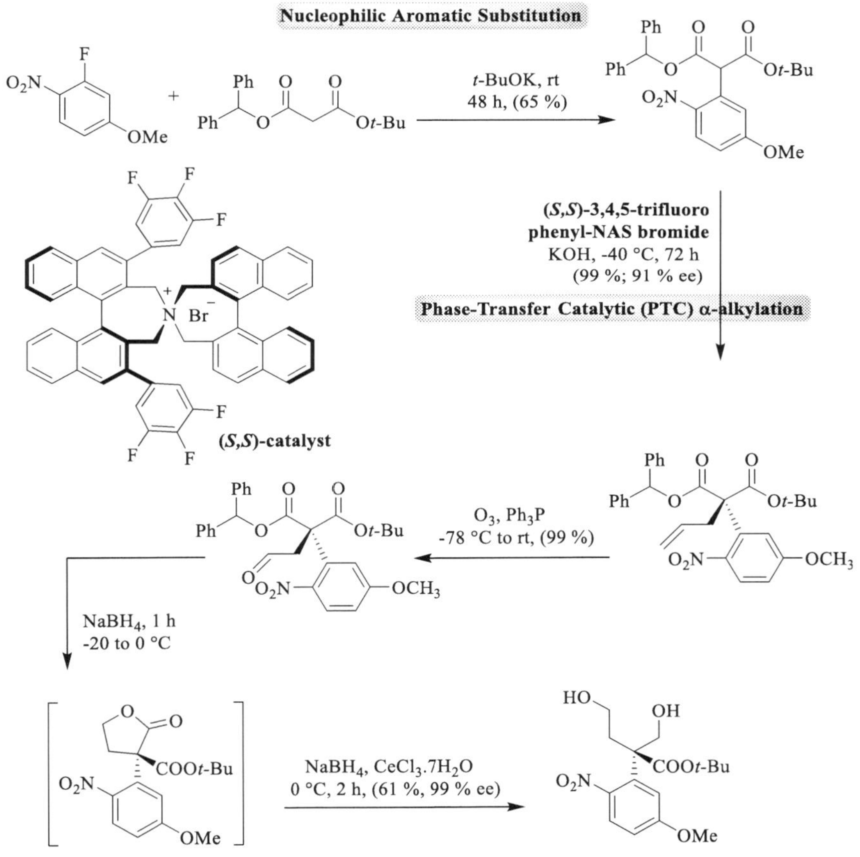

SCHEME 6.16 Park's enantioselective total synthesis of (−)-horsfiline.[20]

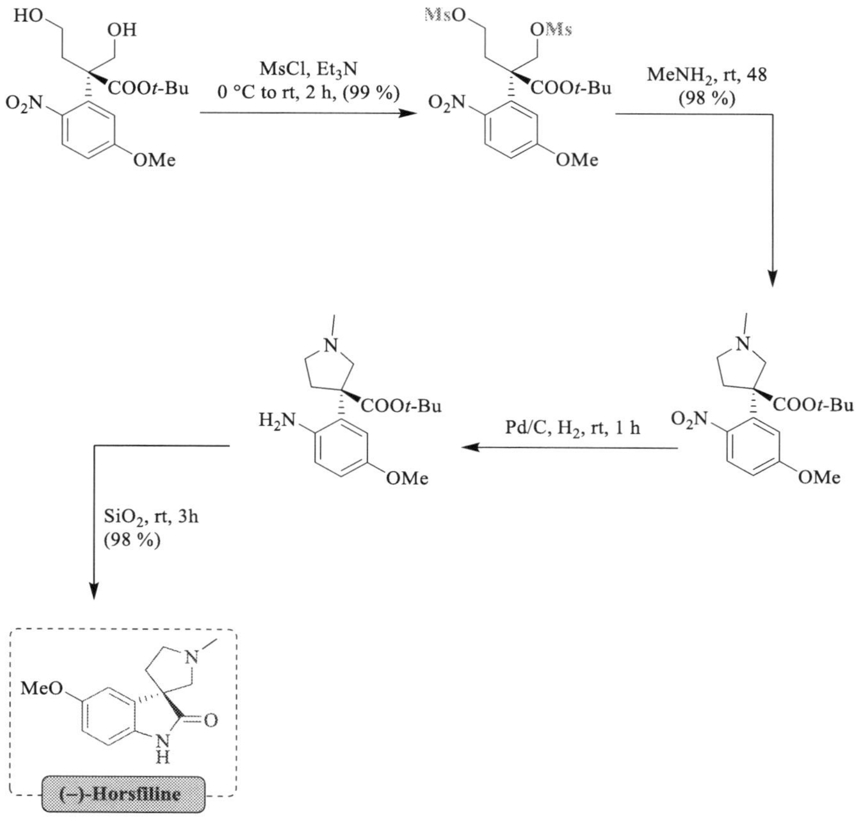

82

6. Coerulescine and horsfiline

SCHEME 6.16 (Continued).

Key features

- (−)-Horsfiline is obtained from diphenylmethyl *tert*-butyl malonate in 9 steps (32%, >99% *ee*) by using the enantioselective phase-transfer catalytic allylation (91% *ee*) as the key step.
- Preparation of α-arylmalonate via nucleophilic aromatic substitution.
- Asymmetric PTC allylation of α-arylmalonate by catalysis of (*S,S*)-3,4,5-trifluoro phenyl-NAS bromide.
- Intramolecular lactamization.
- Reduction and spirocyclization.

6.1.17 Hayashi's enantioselective total synthesis of (−)-coerulescine and (−)-horsfiline (2014)

See Scheme 6.17.

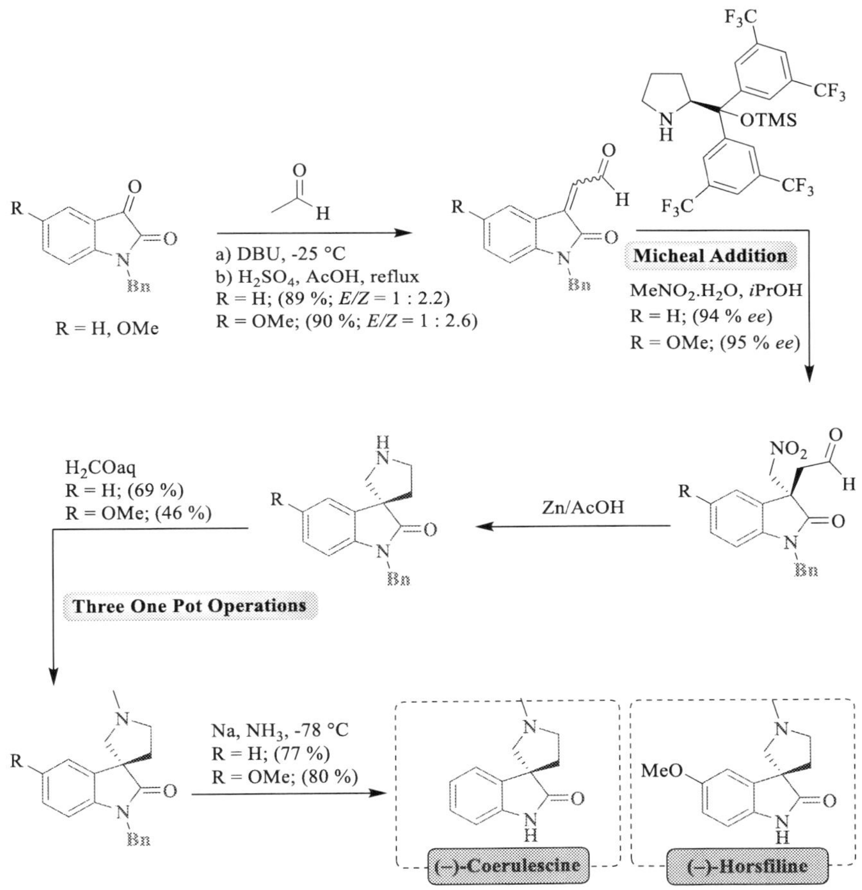

SCHEME 6.17   Hayashi's total synthesis of (−)-coerulescine and (−)-horsfiline.[21]

Key features

- (−)-Coerulescine and (−)-horsfiline were synthesized through three one-pot operations in 46% and 33% overall yields, respectively.
- Synthesis of 2-oxoindoline-3-ylidene acetaldehydes from isatin derivatives via Aldol reactions.
- Asymmetric Michael addition of nitromethane to a 2-oxoindoline-3-ylidene acetaldehyde catalyzed by diarylprolinol silyl ether.

84

6. Coerulescine and horsfiline

- Spirocyclization to produce carbon-stereocenter.
- In this approach a mixture of *E* and *Z* isomers are used as precursors giving the enantioselective product in best yields. This is a synthetic advantage, because there is no need of pure *Z* isomer as precursor.

6.1.18 Castillo's formal synthesis of (±)-coerulescine (2009)

See Scheme 6.18.

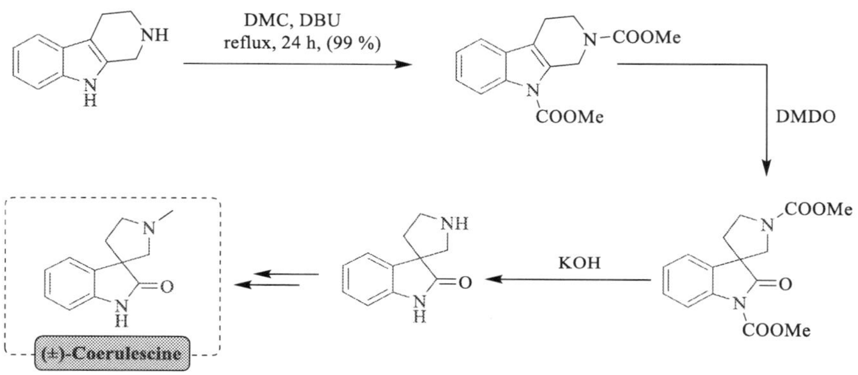

SCHEME 6.18 Castillo's formal synthesis of (±)-coerulescine.[22]

Key features

- DMDO mediated formal synthesis of (±)-coerulescine
- DMDO performs oxidative rearrangement of β-carbolines having electron-withdrawing groups at both *N*2 and *N*9 positions

References

1. Jossang, A.; Jossang, P.; Hadi, H. A.; Sevenet, T.; Bodo, B. Horsfiline, An Oxindole Alkaloid From *Horsfieldia superba*. *J. Org. Chem.* **1991**, *56* (23), 6527–6530.
2. Bascop, S.-I.; Sapi, J.; Laronze, J.-Y.; Levy, J. On the Synthesis of the Oxindole Alkaloid: (±)-Horsfiline. *Heterocycles (Sendai)* **1994**, *38* (4), 725–732.
3. Kuehne, M. E.; Roland, D. M.; Hafter, R. Studies in Biomimetic Alkaloid Syntheses. 2. Synthesis of Vincadifformine From Tetrahydro-. Beta.-Carboline Through a Secodine Intermediate. *J. Org. Chem.* **1978**, *43* (19), 3705–3710.
4. Anderton, N.; Cockrum, P. A.; Colegate, S. M.; Edgar, J. A.; Flower, K.; Vit, I.; Willing, R. I. Oxindoles From *Phalaris coerulescens*. *Phytochemistry* **1998**, *48* (3), 437–439.
5. Jones, K.; Wilkinson, J. A Total Synthesis of Horsfiline via Aryl Radical Cyclisation. *J. Chem. Soc. Chem. Commun.* **1992**, *24*, 1767–1769.
6. Lakshmaiah, G.; Kawabata, T.; Shang, M.; Fuji, K. Total Synthesis of (−)-Horsfiline via Asymmetric Nitroolefination. *J. Org. Chem.* **1999**, *64* (5), 1699–1704.

Indole Alkaloids

References

7. Fischer, C.; Meyers, C.; Carreira, E. M. Efficient Synthesis of (±)-Horsfiline Through the MgI2-Catalyzed Ring-Expansion Reaction of a Spiro[cyclopropane-1,3'-indol]-2'-one. *Helv. Chim. Acta* **2000**, *83* (6), 1175–1181.

8. Somei, M.; Noguchi, K.; Yamagami, R.; Kawada, Y.; Yamada, K.; Yamada, F. Preparation and a Novel Rearrangement Reaction of 1,2,3,4-Tetrahydro-9-hydroxy-3-carboline, and Their Applications for the Total Synthesis of (±)-Coerulescine. *Heterocycles* **2000**, *53* (1), 7–10.

9. Cravotto, G.; Giovenzana, G. B.; Pilati, T.; Sisti, M.; Palmisano, G. Azomethine Ylide Cycloaddition/Reductive Heterocyclization Approach to Oxindole Alkaloids: Asymmetric Synthesis of (−)-Horsfiline. *J. Org. Chem.* **2001**, *66* (25), 8447–8453.

10. Selvakumar, N.; Azhagan, A. M.; Srinivas, D.; Krishna, G. G. A Direct Synthesis of 2-Arylpropenoic Acid Esters Having Nitro Groups in the Aromatic Ring: A Short Synthesis of (±)-Coerulescine and (±)-Horsfiline. *Tetrahedron Lett.* **2002**, *43* (50), 9175–9178.

11. Lizos, D. E.; Murphy, J. A. Concise Synthesis of (±)-Horsfiline and (±)-Coerulescine by Tandem Cyclisation of Iodoaryl Alkenyl Azides. *Org. Biomol. Chem.* **2003**, *1* (1), 117–122.

12. Chang, M.-Y.; Pai, C.-L.; Kung, Y.-H. Synthesis of (±)-Coerulescine and a Formal Synthesis of (±)-Horsfiline. *Tetrahedron Lett.* **2005**, *46* (49), 8463–8465.

13. Trost, B. M.; Brennan, M. K. Palladium Asymmetric Allylic Alkylation of Prochiral Nucleophiles: Horsfiline. *Org. Lett.* **2006**, *8* (10), 2027–2030.

14. Jaegli, S.; Vors, J.-P.; Neuville, L.; Zhu, J. Total Synthesis of Horsfiline: A Palladium-Catalyzed Domino Heck-Cyanation Strategy. *Synlett* **2009**, *2009* (18), 2997–2999.

15. Deppermann, N.; Thomanek, H.; Prenzel, A. H.; Maison, W. Pd-Catalyzed Assembly of Spirooxindole Natural Products: A Short Synthesis of Horsfiline. *J. Org. Chem.* **2010**, *75* (17), 5994–6000.

16. Kulkarni, A. A. Continuous Flow Nitration in Miniaturized Devices. *Beilstein J. Org. Chem.* **2014**, *10* (1), 405–424.

17. White, J. D.; Li, Y.; Ihle, D. C. Tandem Intramolecular Photocycloaddition–Retro-Mannich Fragmentation as a Route to Spiro[pyrrolidine-3,3'-oxindoles] Total Synthesis of (±)-Coerulescine, (±)-Horsfiline, (±)-Elacomine, and (±)-6-Deoxyelacomine. *J. Org. Chem.* **2010**, *75* (11), 3569–3577.

18. Kim, M.-H.; Kim, G.-C. Intramolecular Allylation to Coerulescine and a New Route to Formal Synthesis of Horsfiline. *Bull. Korean Chem. Soc.* **2012**, *33* (6), 1821–1822.

19. Görmen, M.; Le Goff, R.; Lawson, A. M.; Daïch, A.; Comesse, S. Tandem Aza-Michael/Spiro-Ring Closure Sequence: Access to a Versatile Scaffold and Total Synthesis of (±)-Coerulescine. *Tetrahedron Lett.* **2013**, *54* (17), 2174–2176.

20. Hong, S.; Jung, M.; Park, Y.; Ha, M. W.; Park, C.; Lee, M.; Park, H.-g Efficient Enantioselective Total Synthesis of (−)-Horsfiline. *Chem. Eur. J.* **2013**, *19* (29), 9599–9605.

21. Mukaiyama, T.; Ogata, K.; Sato, I.; Hayashi, Y. Asymmetric Organocatalyzed Michael Addition of Nitromethane to a 2-Oxoindoline-3-ylidene Acetaldehyde and the Three One-Pot Sequential Synthesis of (−)-Horsfiline and (−)-Coerulescine. *Chem. Eur. J.* **2014**, *20* (42), 13583–13588.

22. Suárez-Castillo, O. R.; Meléndez-Rodríguez, M.; Contreras-Martínez, Y. M.; Álvarez-Hernández, A.; Morales-Ríos, M. S.; Joseph-Nathan, P. DMD Mediated Formal Synthesis of (±)-Coerulescine. *Nat. Prod. Commun.* **2009**, *4* (6). 1934578X0900400612.

CHAPTER 7

Elacomine and isoelacomine

Elacomine was first isolated by Slywka in 1969 from a shrub *Elaeagnus commutata*.[1] Elacomine and isoelacomine (Fig. 7.1) both occur naturally in racemic form as elacomine can easily be isomerized to isoelacomine. None of the biological activity is associated with any of these molecules, but still, these are tried much to be synthesized in lab due to their simple spiro(pyrrolidine-3,3'-oxindole) structure present in many biologically active natural products.[2]

Different synthetic strategies have been applied in lab synthesis starting from tryptamine including oxidative rearrangement of β-carbolines, intramolecular spirocyclization of iminium ion, tandem intramolecular photocycloaddition, and retro-Mannich fragmentation. Up to now, a total of three racemic and only one asymmetric synthesis are reported for (+)-elacomine.

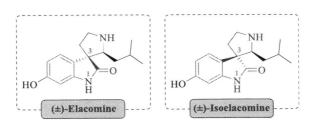

FIGURE 7.1   Structures of (+)-elacomine and (−)-isoelacomine.

88

7. Elacomine and isoelacomine

7.1 Total syntheses of (±)-elacomine and (±)-isoelacomine

7.1.1 Borschberg's racemic total synthesis of (+)-elacomine and (−)-isoelacomine (1996)

See Schemes 7.1 and 7.2.

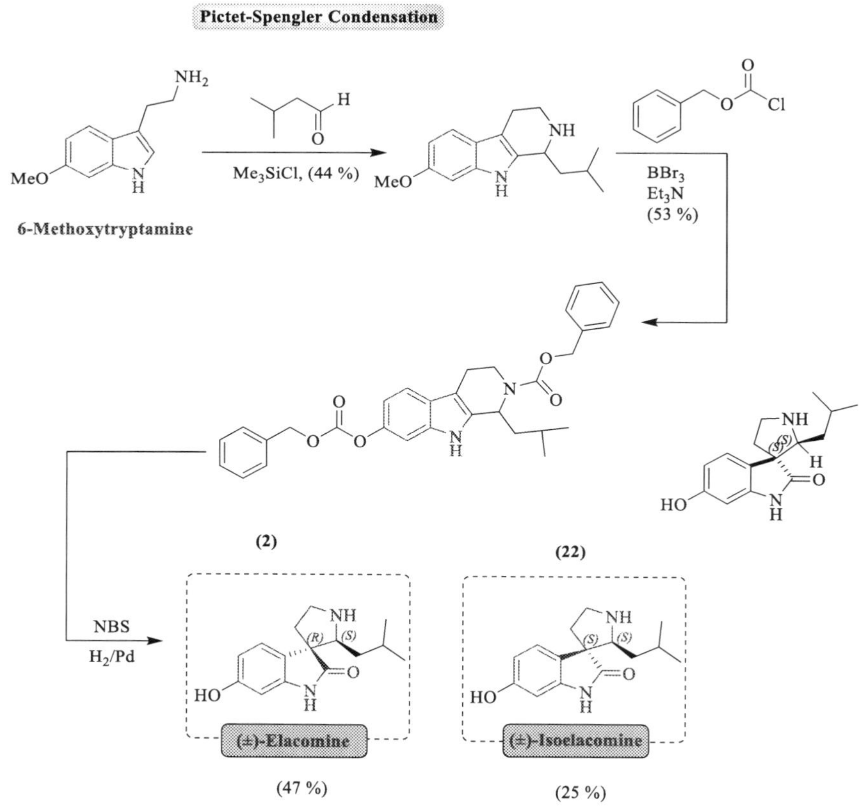

SCHEME 7.1 Borschberg's total synthesis of (±)-elacomine and (±)-isoelacomine.[3]

Indole Alkaloids

7.1 Total syntheses of (±)-elacomine and (±)-isoelacomine

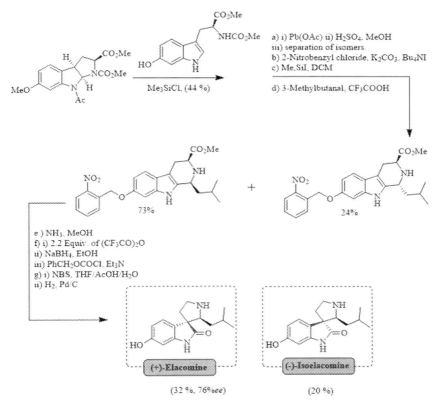

SCHEME 7.2 Borschberg's total synthesis of (+)-elacomine and (−)-isoelacomine.[3]

Key features

- Pictet–Spengler condensation between 6-methoxytryptamine and isovaleraldehyde to produce β-carboline
- Ether cleavage and double protection of produced carboline
- Oxidative rearrangements
- Hydrogenolysis deprotection

90
7. Elacomine and isoelacomine

7.1.2 Horne's racemic total synthesis of (±)-elacomine and (±)-isoelacomine (2004)

See Scheme 7.3.

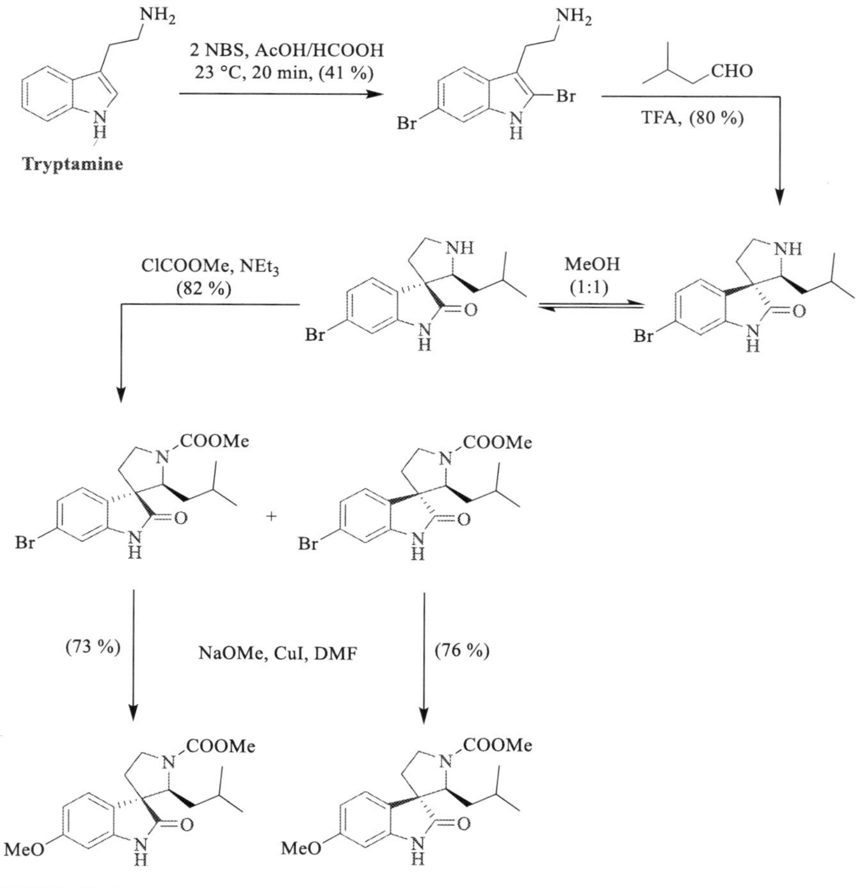

SCHEME 7.3 Horne's total synthesis of (±)-elacomine and (±)-isoelacomine.[4]

Indole Alkaloids

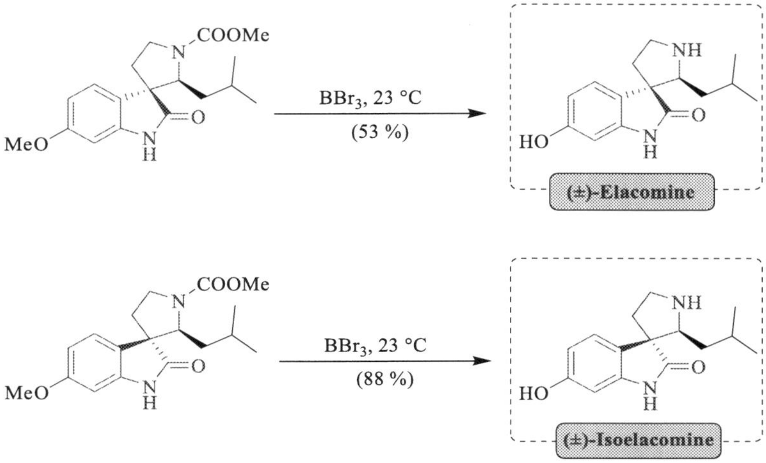

7.1 Total syntheses of (±)-elacomine and (±)-isoelacomine

SCHEME 7.3 (Continued).

Key features

- Preparation of 2-halotryptamines from tryptamine
- Stereoselective intramolecular iminium ion spirocyclization methodology for the synthesis of elacomine and isoelacomine

92

7.1.3 White's racemic total synthesis of (±)-elacomine (2010)

See Scheme 7.4.

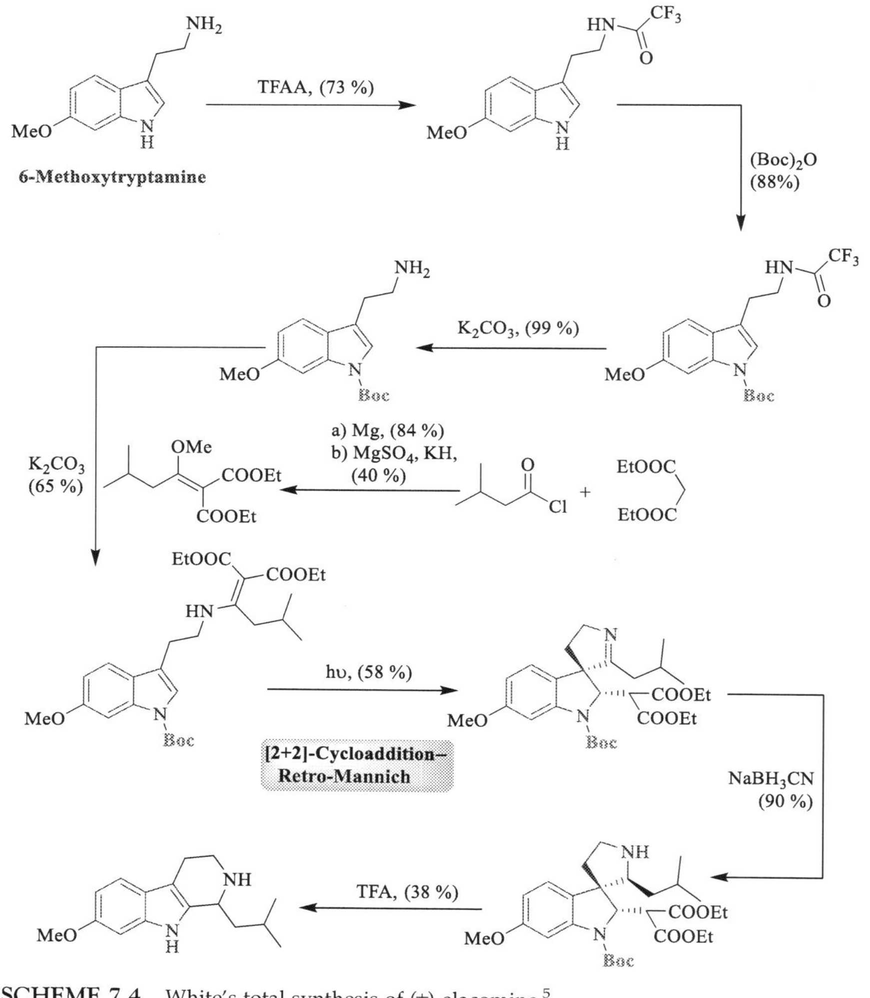

SCHEME 7.4 White's total synthesis of (±)-elacomine.[5]

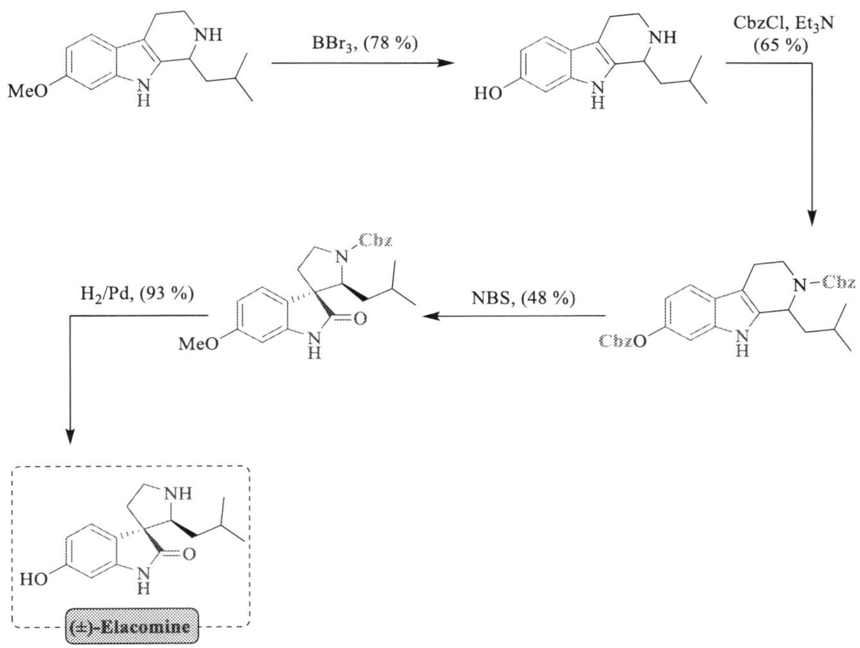

7.1 Total syntheses of (±)-elacomine and (±)-isoelacomine 93

SCHEME 7.4 (Continued).

Key features

- Condensation of protected 6-methoxytryptamine with β-ethoxymethylidenemalonate to yield photo substrate
- Intramolecular [2 + 2] photocycloaddition and in situ retro-Mannich fragmentation to yield spiropyrrolidine
- Second spontaneous retro-Mannich fragmentation and oxidative rearrangement

94

7.1.4 Njardarson's asymmetric total synthesis of (+)-elacomine (2015)

See Scheme 7.5.

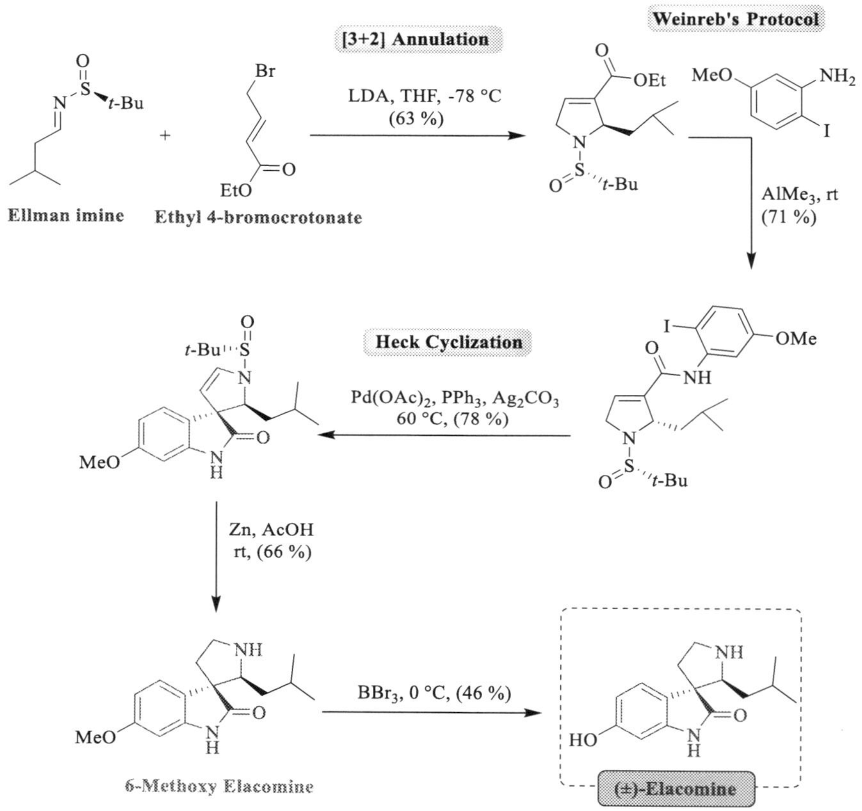

SCHEME 7.5 Njardarson's asymmetric total synthesis of (+)-elacomine.[6]

Key features

- Shortest asymmetric total synthesis of (+)-elacomine in only six steps from isovaleraldehyde
- [3 + 2] Annulation of Ellman imine of isovaleraldehyde with the enolate of ethyl 4-bromocrotonate
- Conversion of resultant pyrroline ester into iodo arylamide using Weinreb's protocol
- Substrate controlled Heck cyclization to form the spirooxindole core
- Reduction of imine to yield methoxy elacomine and standard deprotection leading to (+)-elacomine

7.1.5 Takemoto's formal synthesis of elacomine and isoelacomine (2011)

See Scheme 7.6.

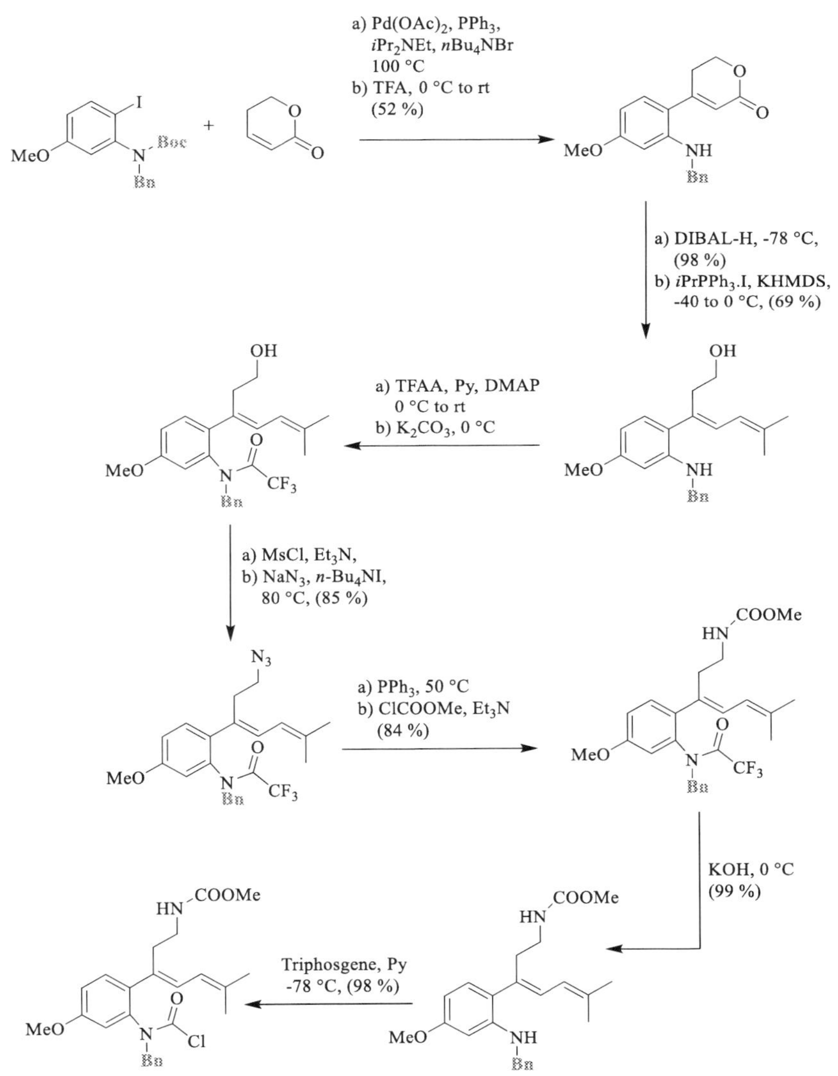

SCHEME 7.6 Takemoto's formal synthesis of (±)-elacomine and (±)-isoelacomine.[7]

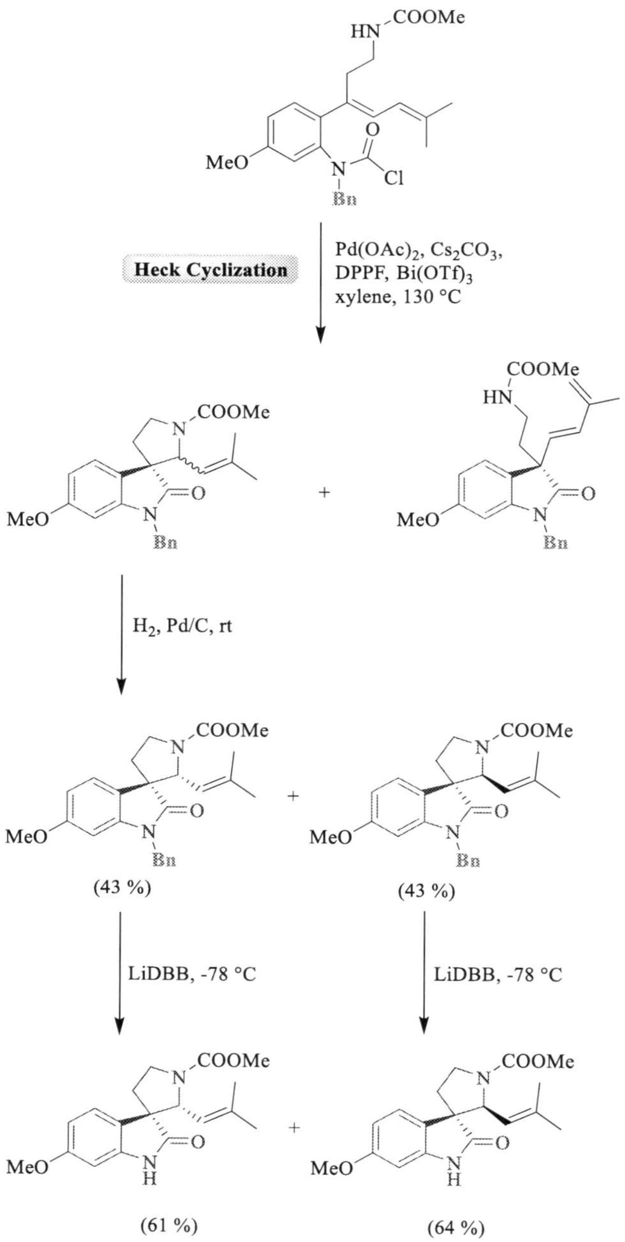

7. Elacomine and isoelacomine

SCHEME 7.6 (Continued).

Indole Alkaloids

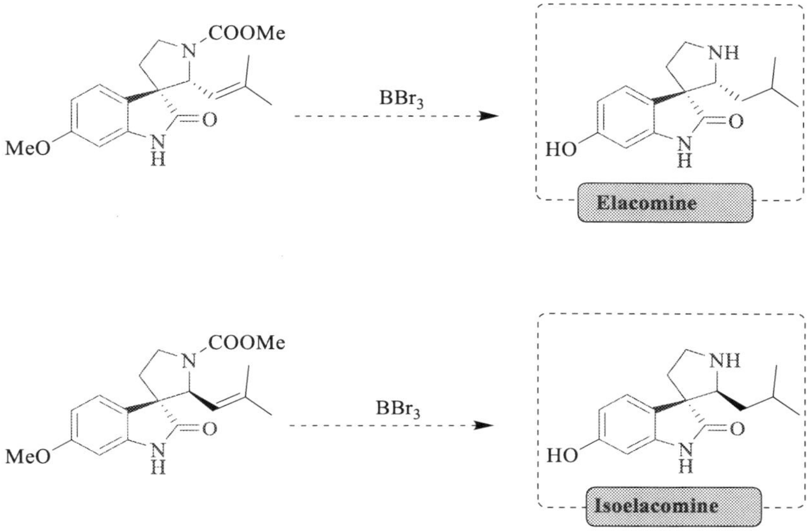

SCHEME 7.6 (Continued).

Key features

- Condensation of iodoaniline derivative with commercially available 5,6-dihydro-2*H*-pyran-2-one
- Conversion of coupling product to carbamoyl chloride with an incorporated diene unit
- Domino palladium-catalyzed Heck reaction and bismuth catalyzed hydroamination to afford spiro product

References

1. Slywka, G. W. A. *Structure of a new beta-carboline alkaloid from* Elaeagnus commutata *(silverberry or wolf willow)*; The University of Alberta: Edmonton, 1969.
2. Marti, C.; Carreira, E. M. Construction of Spiro[pyrrolidine-3,3'-oxindoles] — Recent Applications to the Synthesis of Oxindole Alkaloids. *Eur. J. Org. Chem.* **2003,** *2003* (12), 2209–2219.
3. Pellegrini, C.; Weber, M.; Borschberg, H. J. Total Synthesis of (+)-Elacomine and (−)-Isoelacomine, Two Hitherto Unnamed Oxindole Alkaloids From *Elaeagnus commutata. Helv. Chim. Acta* **1996,** *79* (1), 151–168.
4. Miyake, F. Y.; Yakushijin, K.; Horne, D. A. Preparation and Synthetic Applications of 2-Halotryptamines: Synthesis of Elacomine and Isoelacomine. *Org. Lett.* **2004,** *6* (5), 711–713.
5. White, J. D.; Li, Y.; Ihle, D. C. Tandem intramolecular Photocycloaddition — Retro-Mannich Fragmentation as a Route to Spiro[pyrrolidine-3,3'-oxindoles]. Total Synthesis of (±)-Coerulescine, (±)-Horsfiline, (±)-Elacomine, and (±)-6-Deoxyelacomine. *J. Org. Chem.* **2010,** *75* (11), 3569–3577.

6. Chogii, I.; Njardarson, J. T. Asymmetric [3 + 2] Annulation Approach to 3-Pyrrolines: Concise Total Syntheses of (−)-Supinidine, (−)-Isoretronecanol, and (+)-Elacomine. *Angew. Chem. Int. Ed.* **2015,** *127* (46), 13910−13914.
7. Kamisaki, H.; Nanjo, T.; Tsukano, C.; Takemoto, Y. Domino Pd-Catalyzed Heck Cyclization and Bismuth-Catalyzed Hydroamination: Formal Synthesis of Elacomine and Isoelacomine. *Chem. Eur. J.* **2011,** *17* (2), 626−633.

CHAPTER 8

Gelsemine

Gelsemine (Fig. 8.1) was first isolated in 1870 by Wormley from *Gelsemium sempervirens*. It was an impure alkaloidal product. Six years later, the primary component of this impure alkaloidal product was discovered by Sonnenschein as an amorphous base named gelsemine.[1] This alkaloid was isolated as pure crystalline compound in 1910 and Moore[2] corrected its formula as $C_{20}H_{22}N_2O_2$ but no structural elucidation was done at that time. The structure was elucidated eventually in 1959 by Lovell and coworkers using X-ray crystallography.[3] More recently, in the 1980s, gelsemine was isolated from the roots of a plant having Southeast Asian origin; *Gelsemium elegans*.[4] Moreover, trace quantities were also detected in the stems of south-eastern US plant; *Gelsemium rankinii*.[5]

Gelsemine was thought to have no significant biological activities. A fresh report pointed out its potential to act as antinociception thus relieving chronic pain by acting on the three spinal glycine receptors.

Synthetic chemists tried different strategies for asymmetric synthesis of this molecule due to its complex molecular architecture and potential medicinal applications. Up to now, three asymmetric and two racemic

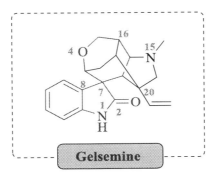

FIGURE 8.1   Structure of gelsemine.

100
8. Gelsemine

syntheses have been reported by different research groups summarized in this section.

8.1 Total syntheses of gelsemine

8.1.1 Fukuyama's first racemic total synthesis of (±)-gelsemine (1996, 1997)

See Scheme 8.1.

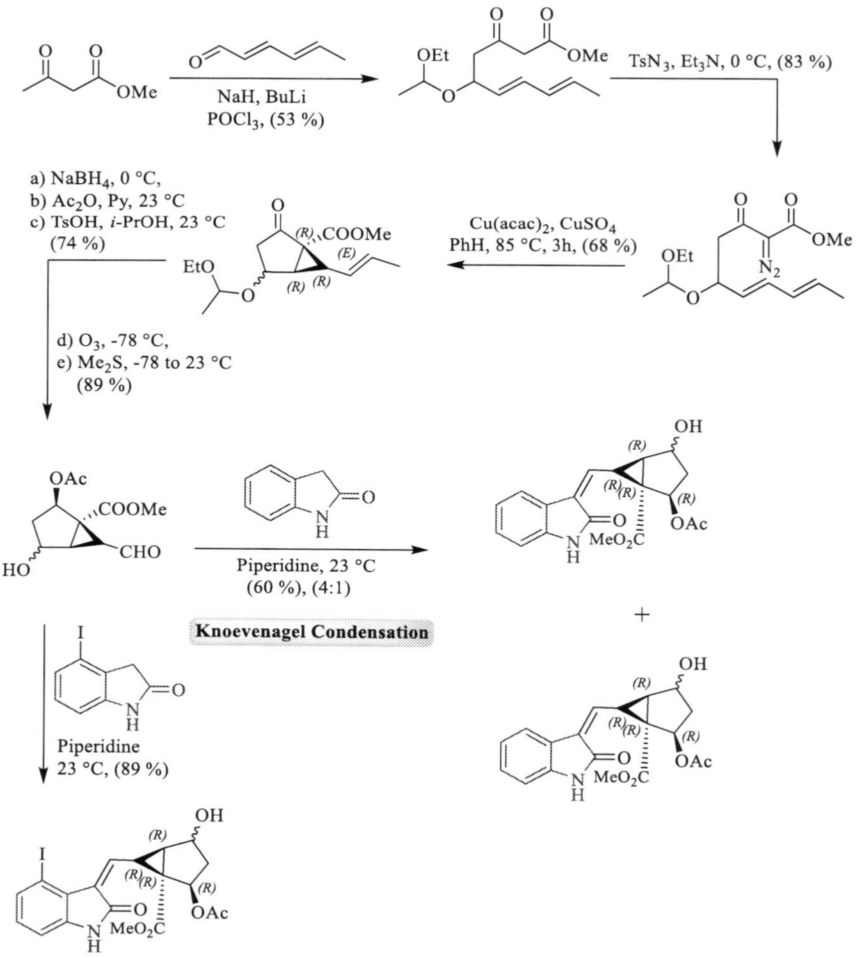

SCHEME 8.1 Fukuyama's first total synthesis of (±)-gelsemine.[6,7]

Indole Alkaloids

8.1 Total syntheses of gelsemine

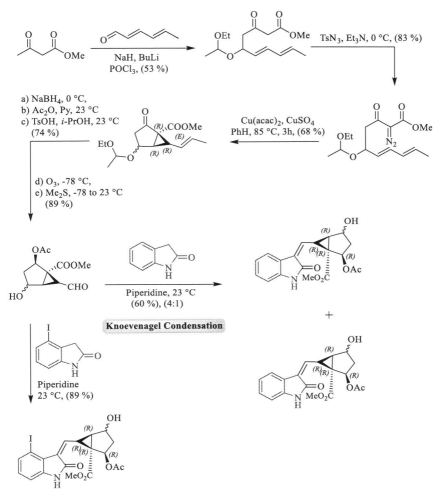

SCHEME 8.1 (Continued).

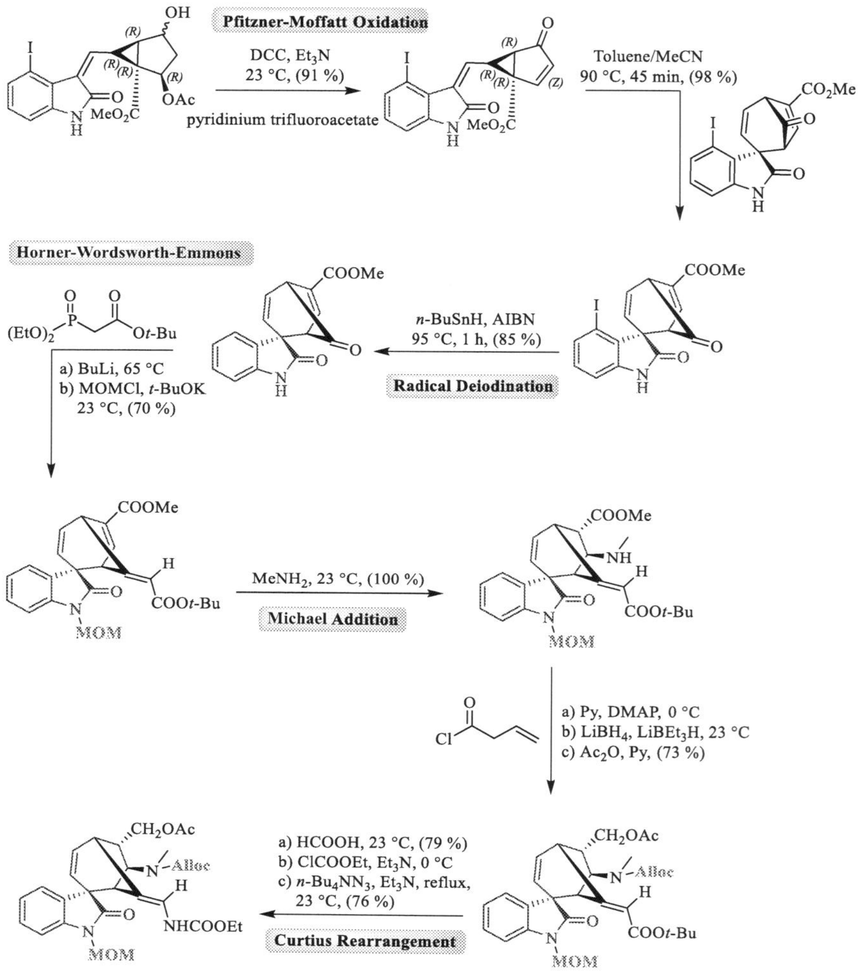

102 8. Gelsemine

SCHEME 8.1 (Continued).

Indole Alkaloids

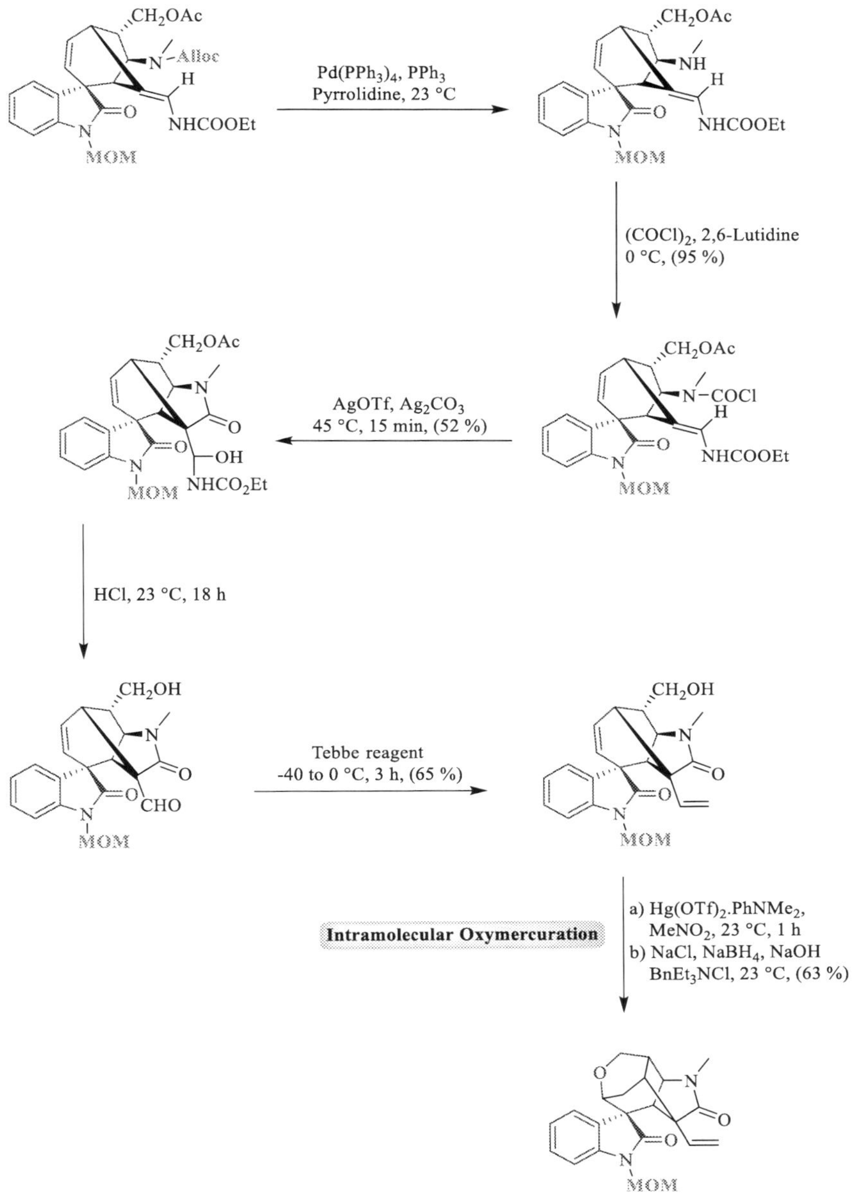

8.1 Total syntheses of gelsemine

103

SCHEME 8.1 (Continued).

Indole Alkaloids

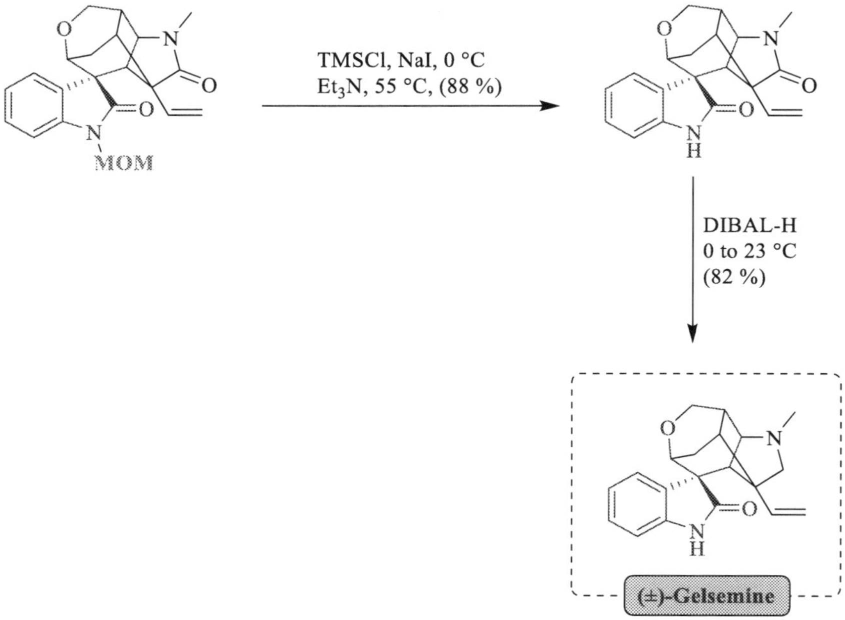

104 8. Gelsemine

SCHEME 8.1 (Continued).

Indole Alkaloids

Key features

- Synthesis of β-keto ester by addition of methyl acetoacetate dianion to sorbic aldehyde followed by immediate protection of unstable alcohol
- Production of Kondo's aldehyde intermediate
- Knoevenagel condensation of aldehyde intermediate with 4-iodooxindole
- Pfitzner–Moffatt oxidation and elimination of acetic acid followed by rearrangement of resultant enone to desired bicyclo[3.2.1] system
- Radical deiodination and synthesis of *tert*-butyl ester
- Michael addition of methyl amine to α,β-unsaturated ester exclusively from the less hindered *exo*-side to afford trans-amino ester
- Protection of amine as allyl carbamate, selective reduction of methyl ester, and acetylation of resultant alcohol
- Conventional Curtius rearrangement to convert *tert*-butyl ester to ethyl urethane
- Conversion to carbamoyl chloride by deprotection and reaction resultant amine with phosgene
- Cyclization to lactam; aminal urethane
- Acetate hydrolysis to give hydroxyl aldehyde and its subsequent methylenation
- Intramolecular oxymercuration according to the Speckamp procedure
- Reduction of resultant organomercurial compound to afford N-MOM-21-oxogelsemine
- Selective reduction of lactam to produce (±)-gelsemine

8.1.2 Fukuyama's enantioselective total synthesis of (+)-gelsemine (2000)

See Scheme 8.2.

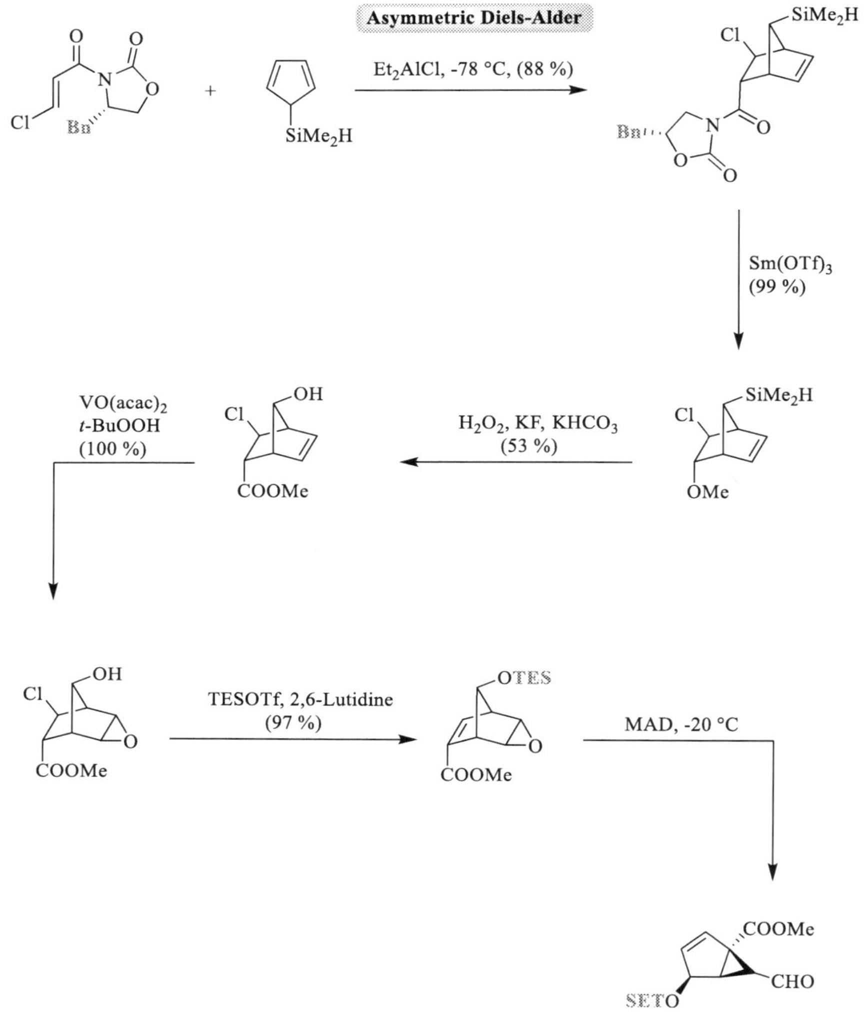

8. Gelsemine

Asymmetric Diels-Alder

SCHEME 8.2 Fukuyama's enantioselective total synthesis of (+)-gelsemine.[8]

8.1 Total syntheses of gelsemine

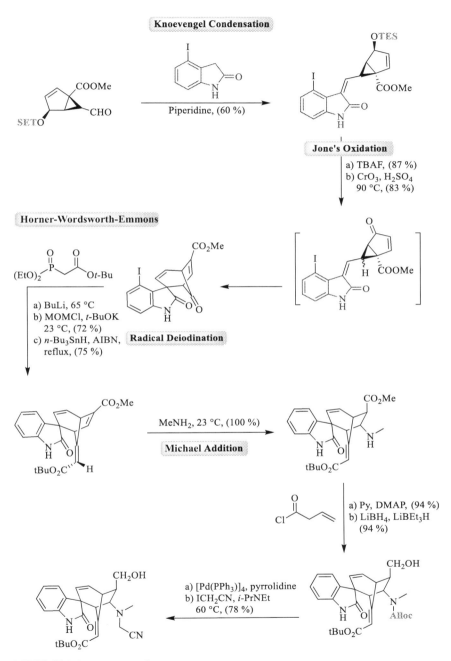

SCHEME 8.2 (Continued).

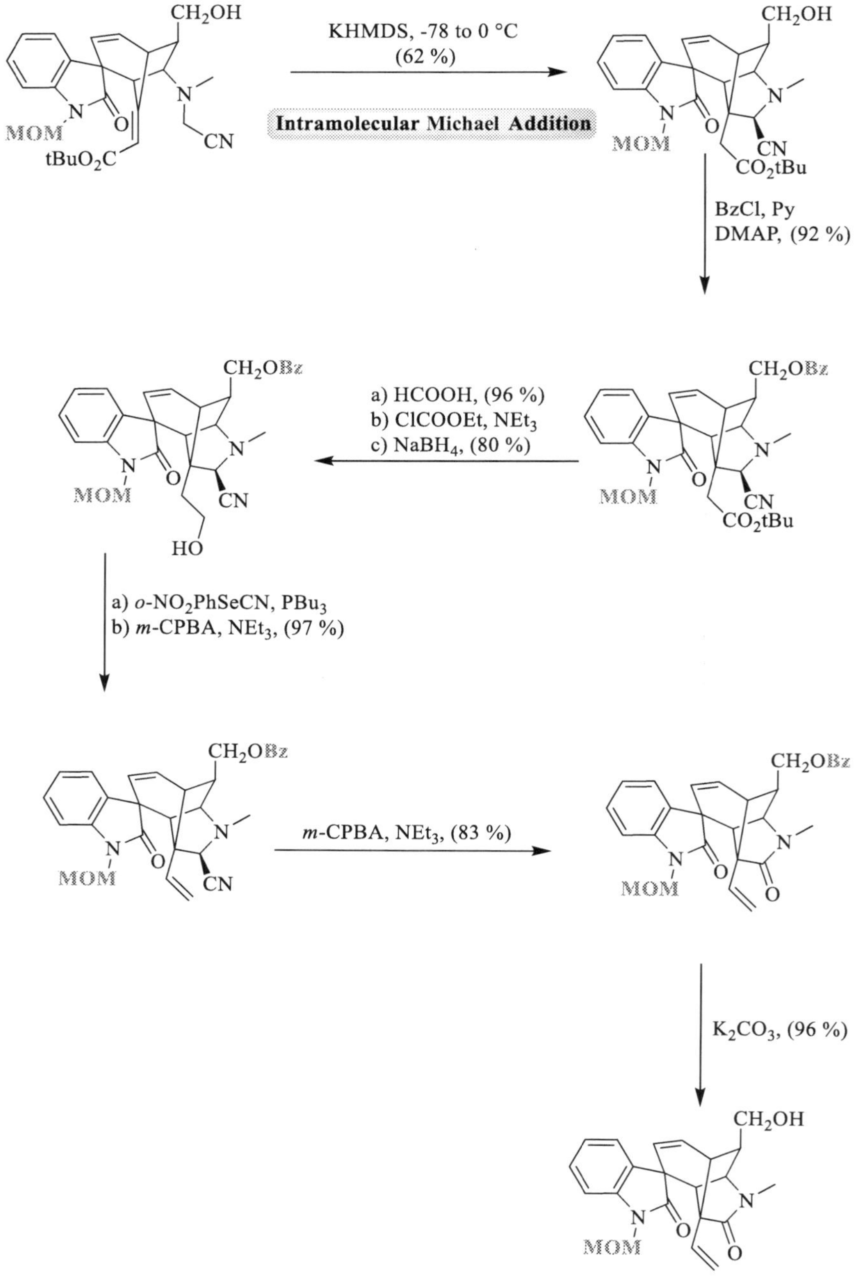

8. Gelsemine

108

SCHEME 8.2 (Continued).

8.1 Total syntheses of gelsemine

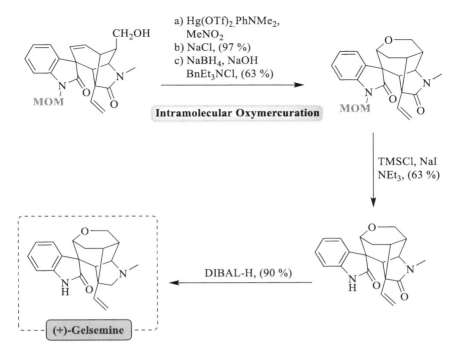

SCHEME 8.2 (Continued).

110

Key features

- Chiral auxiliary controlled asymmetric Diels—Alder reaction
- Oxidation leading to alcohol, epoxidation, protection of alcohol, and dehydrochlorination to yield α,β-unsaturated ester
- Acid-catalyzed rearrangement leading to cyclopropanation
- Knoevenagel condensation of aldehyde intermediate to 4-iodooxindole
- Deprotection followed by Jone's oxidation
- Divinylcyclopropane—cycloheptadiene rearrangement to produce enantiomerically pure bicylco[3.2.1] system
- Horner—Wordsworth—Emmons for elongation of ketone
- One-pot indolinone nitrogen protection and radical deiodination
- Michael addition of methyl amine to strained α,β-unsaturated ester from less hindered *exo*-side leading to *trans*-amino ester
- Allyl carbamate protection to amino group
- Reduction followed by installation of cyanomethyl group on to amine
- Deprotection followed by intramolecular Michael addition
- Protection of alcohol
- Vinyl side chain synthesis according to Grieco's procedure
- Conversion of aminonitrile to lactams
- Oxidation
- Conversion into (+)-gelsemine

8.1.3 Overman's total synthesis of (±)-gelsemine (1999, 2005)

See Scheme 8.3.

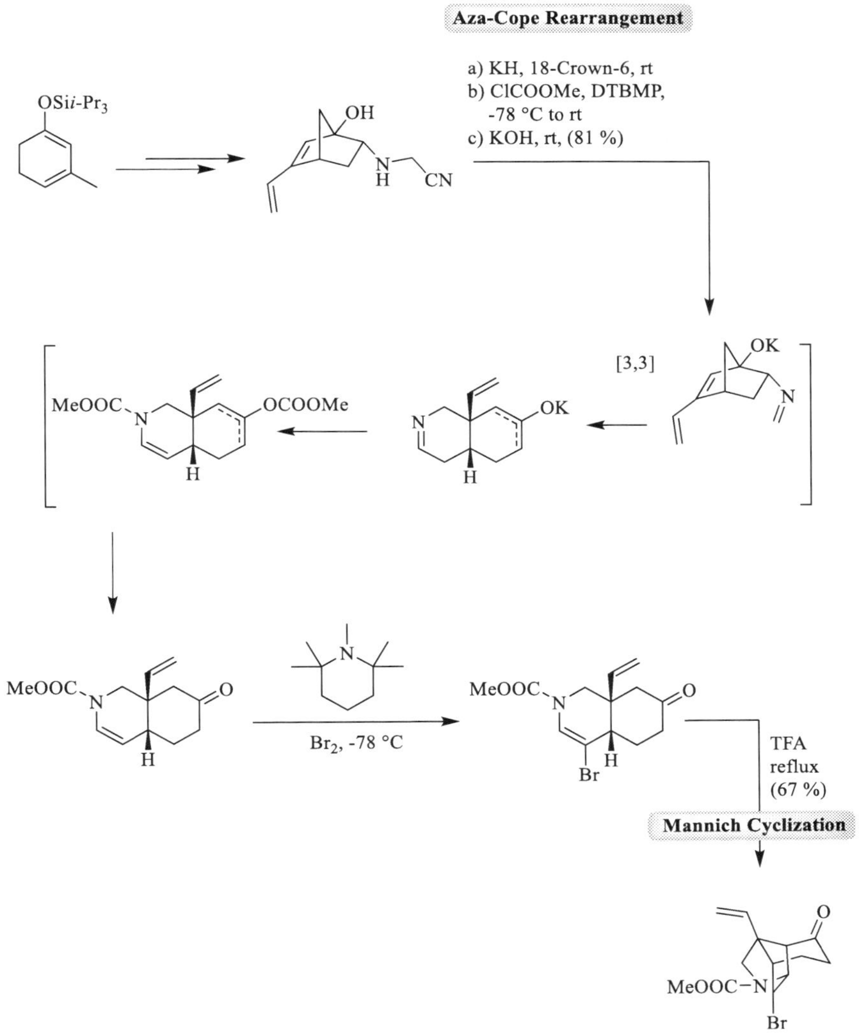
Aza-Cope Rearrangement
a) KH, 18-Crown-6, rt
b) ClCOOMe, DTBMP, -78 °C to rt
c) KOH, rt, (81 %)
OSii-Pr3
OH
N
H
CN
[3,3]
OK
N
MeOOC
N
OCOOMe
H
N
OK
H
MeOOC
N
O
H
N
Br2, -78 °C
MeOOC
N
O
Br
H
TFA
reflux
(67 %)
Mannich Cyclization
O
MeOOC−N
Br

SCHEME 8.3 Overman's total synthesis of (±)-gelsemine.[9,10]

8. Gelsemine

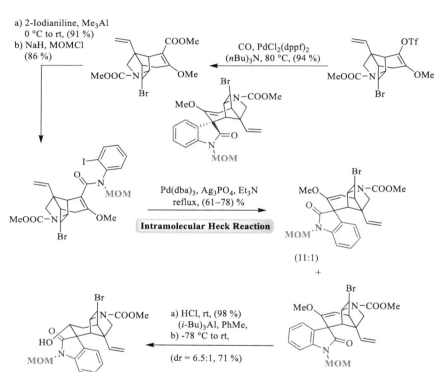

SCHEME 8.3 (Continued).

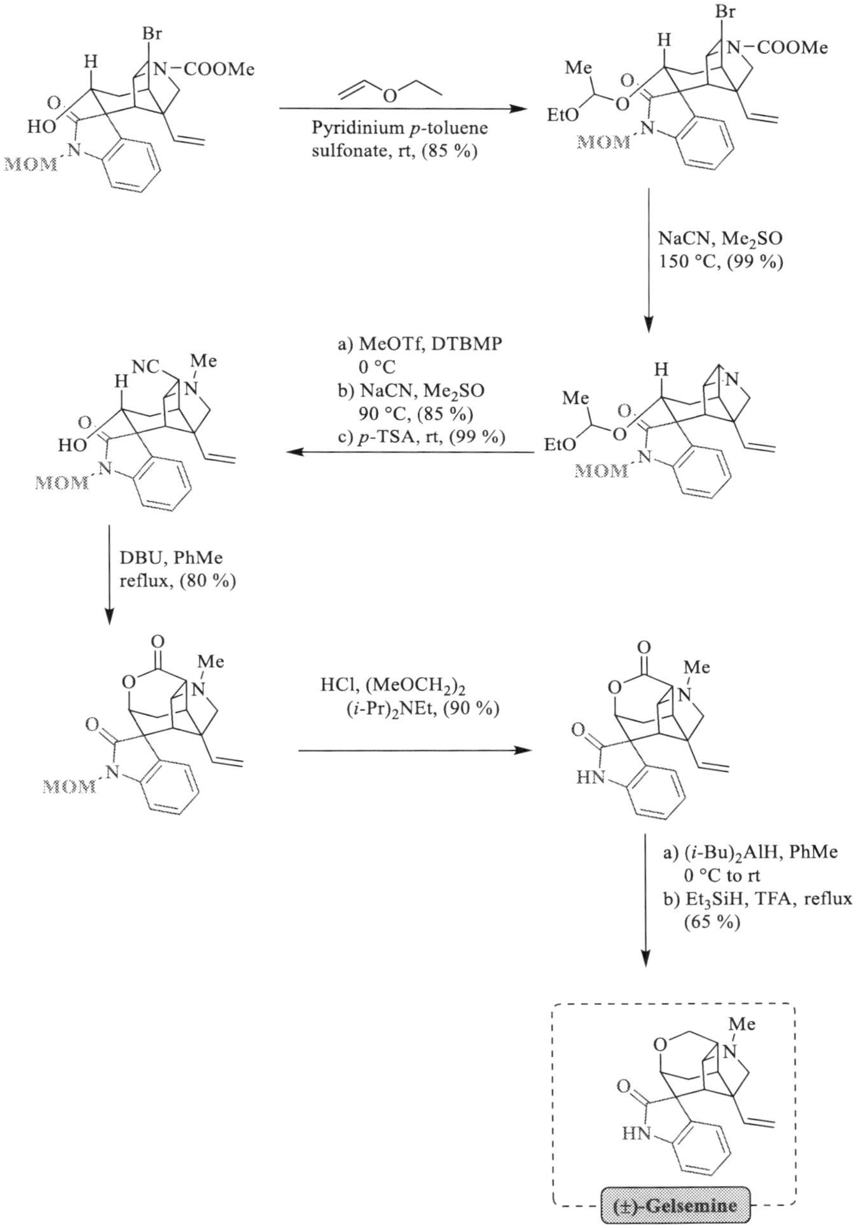

8.1 Total syntheses of gelsemine

113

SCHEME 8.3 (Continued).

Indole Alkaloids

114
8. Gelsemine

Key features

- Conversion of propylsiloxy-3-methyl-1,3-cyclohexadiene into bicyclo [2.2.2]octene in eight steps with 32% overall yield
- 18-Crown-6 mediated anionic aza-Cope rearrangement of derived formaldimine alkoxide
- Conversion of resultant intermediate to *cis*-hexahydroisoquinolinone
- Selective bromination of enecarbamate functional group of *cis*-hexahydroisoquinolinone
- Synthesis of azatricyclodecanone
- Oxidation of enoxytriethylsilane derivative of azatricyclodecanone
- Synthesis of enol triflate via Comin's reagent
- Palladium-catalyzed carbonylation
- Condensation with dimethylaluminum amide of iodoaniline
- Intramolecular Heck reaction to yield spirooxindole stereoisomeric intermediates
- Epimerization of oxindole and completion of remaining oxacyclic ring via complex base-promoted skeletal rearrangement

8.1.4 Qin's total synthesis of (+)-gelsemine (2012)

See Scheme 8.4.

8.1 Total syntheses of gelsemine

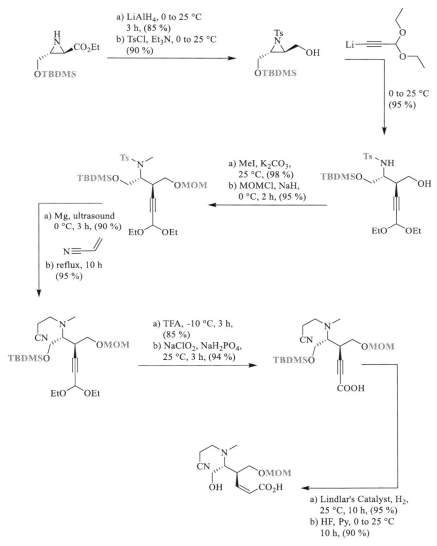

SCHEME 8.4 Qin's total synthesis of (+)-gelsemine.[11]

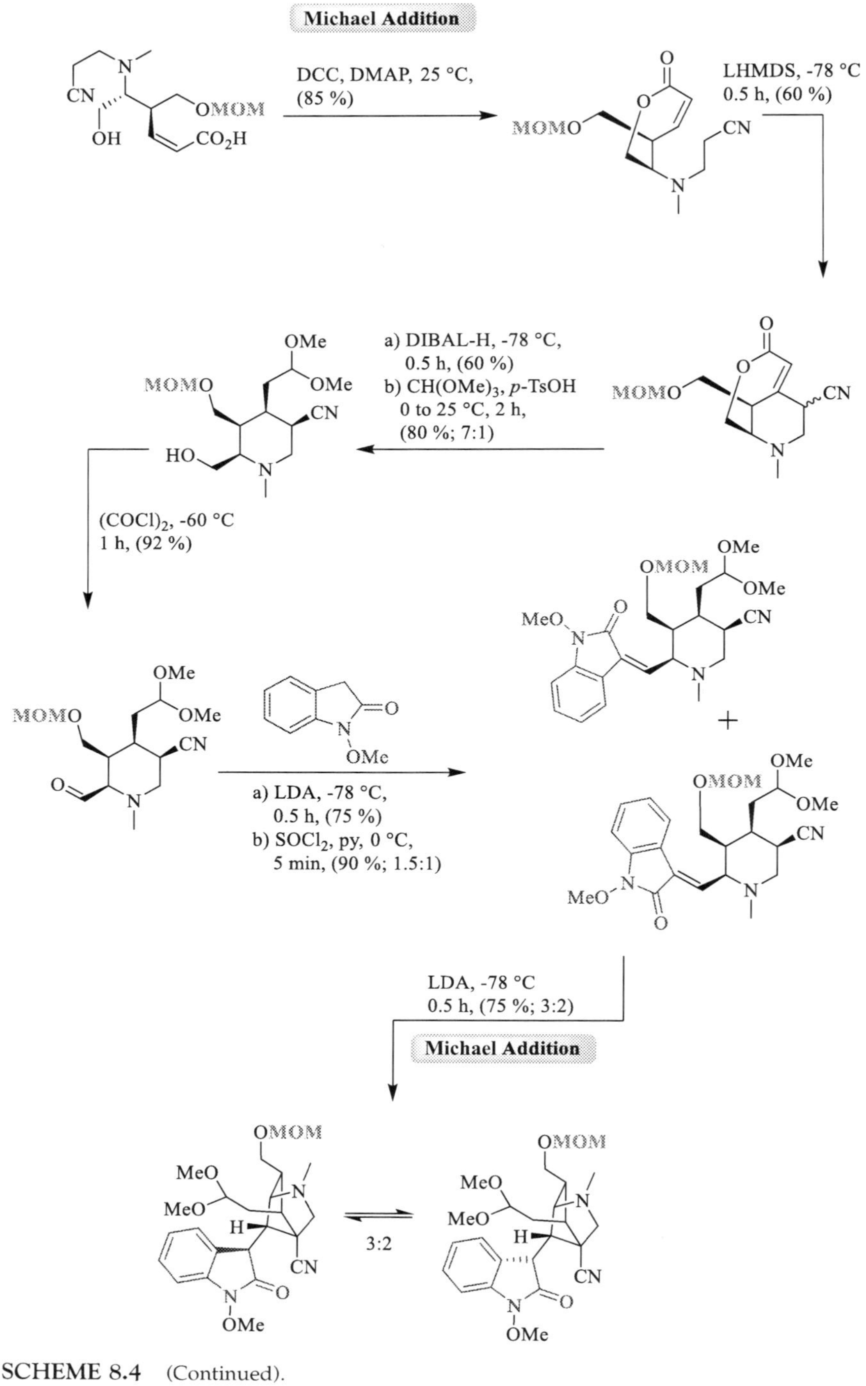

116

8. Gelsemine

Michael Addition

Michael Addition

SCHEME 8.4 (Continued).

Indole Alkaloids

8.1 Total syntheses of gelsemine

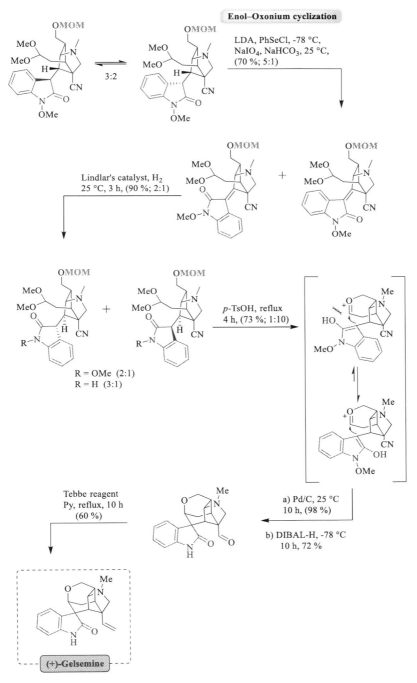

SCHEME 8.4 (Continued).

118

Key features

- Total synthesis of (+)-gelsemine from (*R,R*)-aziridine in 25 steps in 1% overall yield
- Reduction and tosylation of (*R,R*)-aziridine
- Addition of 3,3-diethoxyprop-1-yn-1-yl lithium
- Methylation and protection
- Deprotection and oxidation
- Partial hydrogenation of the alkynyl group
- Generation of oxepinone ring
- Reduction and subsequent aldehyde protection
- Oxidation to produce *cis* tetra-substituted piperidine aldehyde intermediate
- Condensation of produced intermediate with *N*-methoxyoxindole
- Michael addition
- Enol-oxonium cyclization
- (+)-Gelsemine production via hydrogenation and nitrile to alkene conversion

8.1.5 Qiu's asymmetric total synthesis of (+)-gelsemine (2015)

See Scheme 8.5.

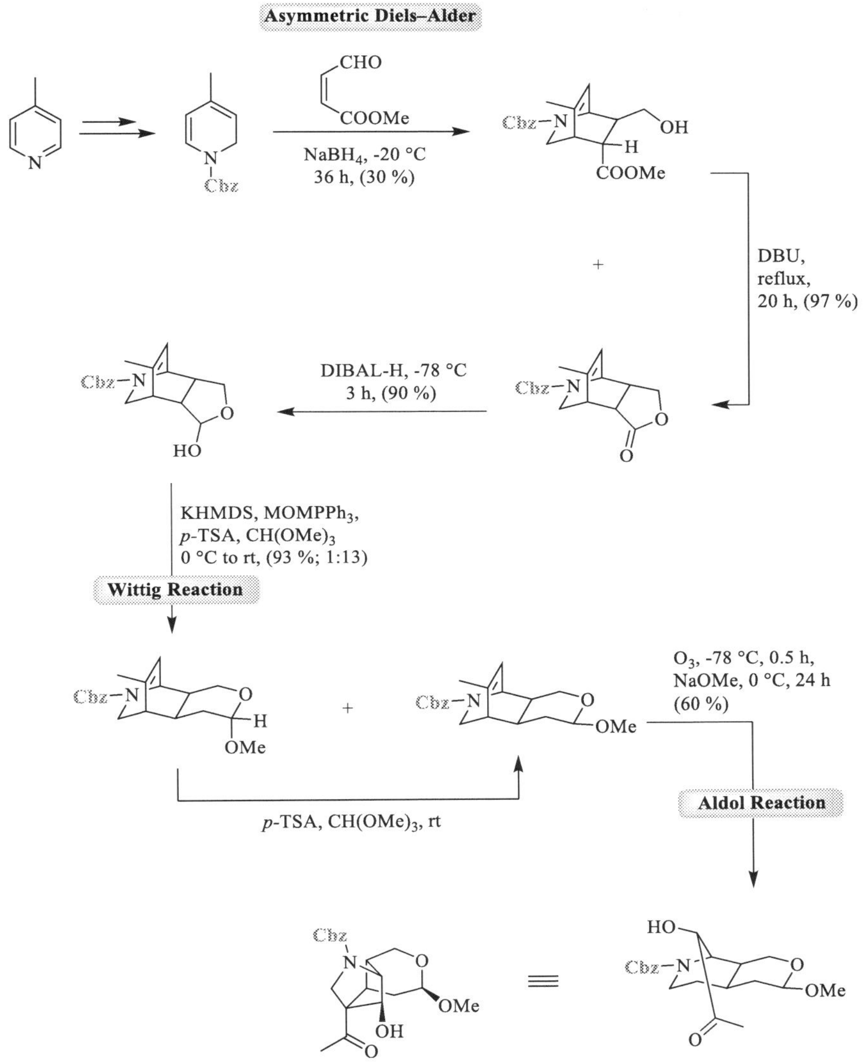

8.1 Total syntheses of gelsemine

119

SCHEME 8.5 Qiu's total synthesis of (+)-gelsemine.[12]

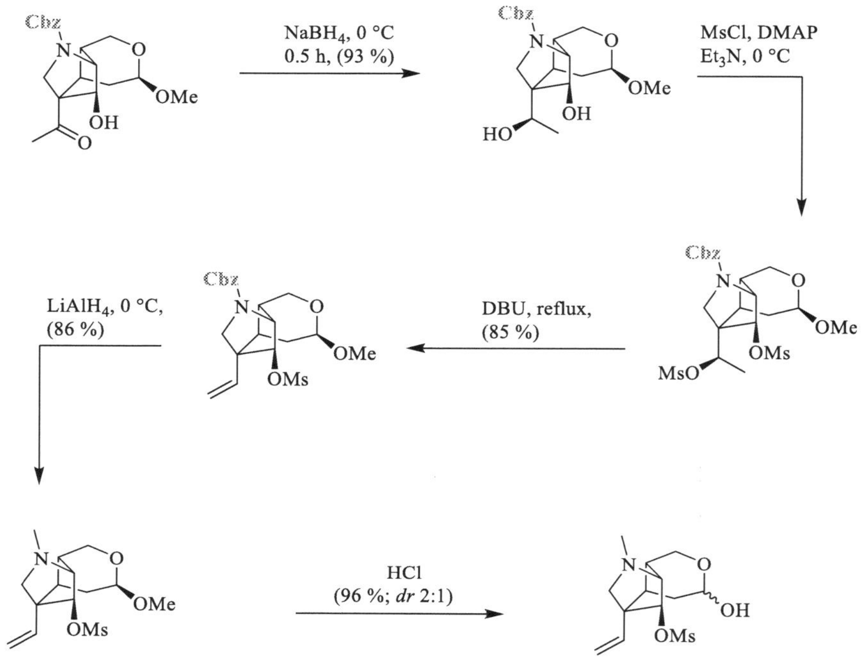

120

8. Gelsemine

SCHEME 8.5 (Continued).

Indole Alkaloids

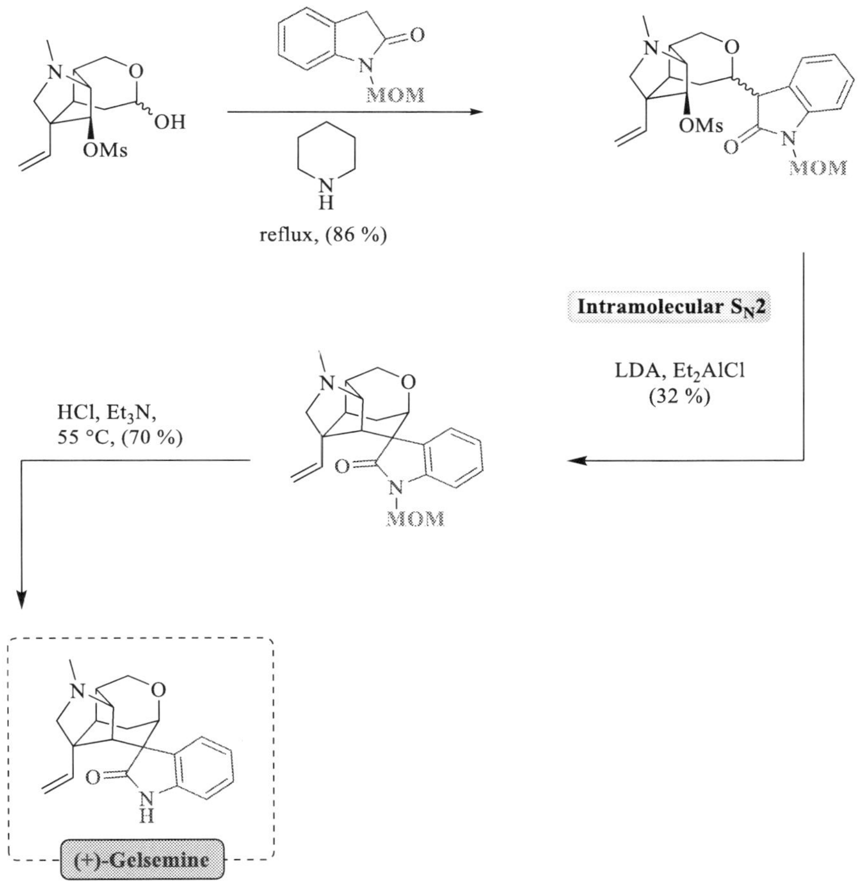

8.1 Total syntheses of gelsemine

121

reflux, (86 %)

Intramolecular S$_N$2

LDA, Et$_2$AlCl
(32 %)

HCl, Et$_3$N,
55 °C, (70 %)

(+)-Gelsemine

SCHEME 8.5 (Continued).

Indole Alkaloids

122

Key features

- (+)-Gelsemine is synthesized in 99% enantiomeric excess and 5% overall yield in 12 steps starting from 4-methyl dihydropyridine that can be prepared from commercially available 4-methylpyridine.
- Asymmetric Diels–Alder cycloaddition.
- Hemiacetal production via selective reduction.
- Wittig reaction and ozonolysis.
- Aldol reaction.
- Reduction of hydroxyketone intermediate to diol.
- Formation of disulfonate.
- Formation of alkene and respective deprotection of nitrogen.
- Acid hydrolysis of acetal to produce hemiacetal.
- Condensation of produced hemiacetal intermediate with methoxymethyl oxindole.
- Intramolecular S_N2 displacement to produce (+)-gelsemine.

8.2 Formal syntheses of gelsemine

8.2.1 Johnson's synthesis of key tetracyclic gelsemine intermediate (1994)

See Scheme 8.6.

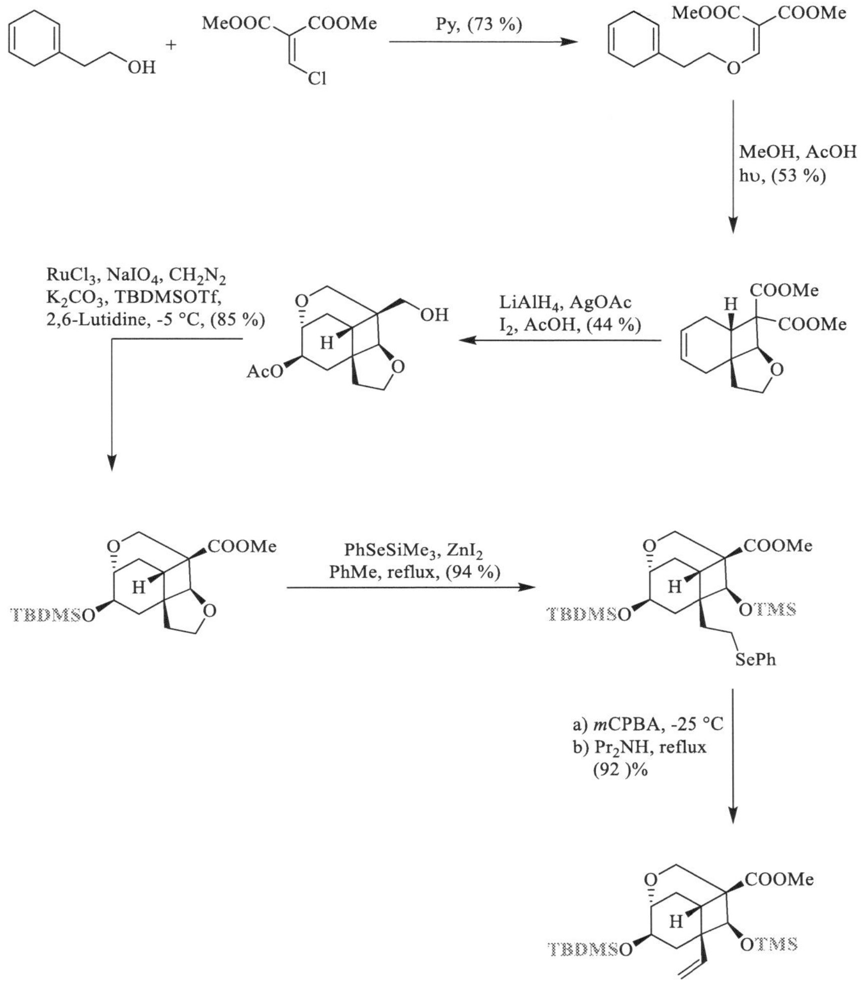

8.2 Formal syntheses of gelsemine

123

SCHEME 8.6 Johnson's synthesis of key tetracyclic gelsemine intermediate.[13]

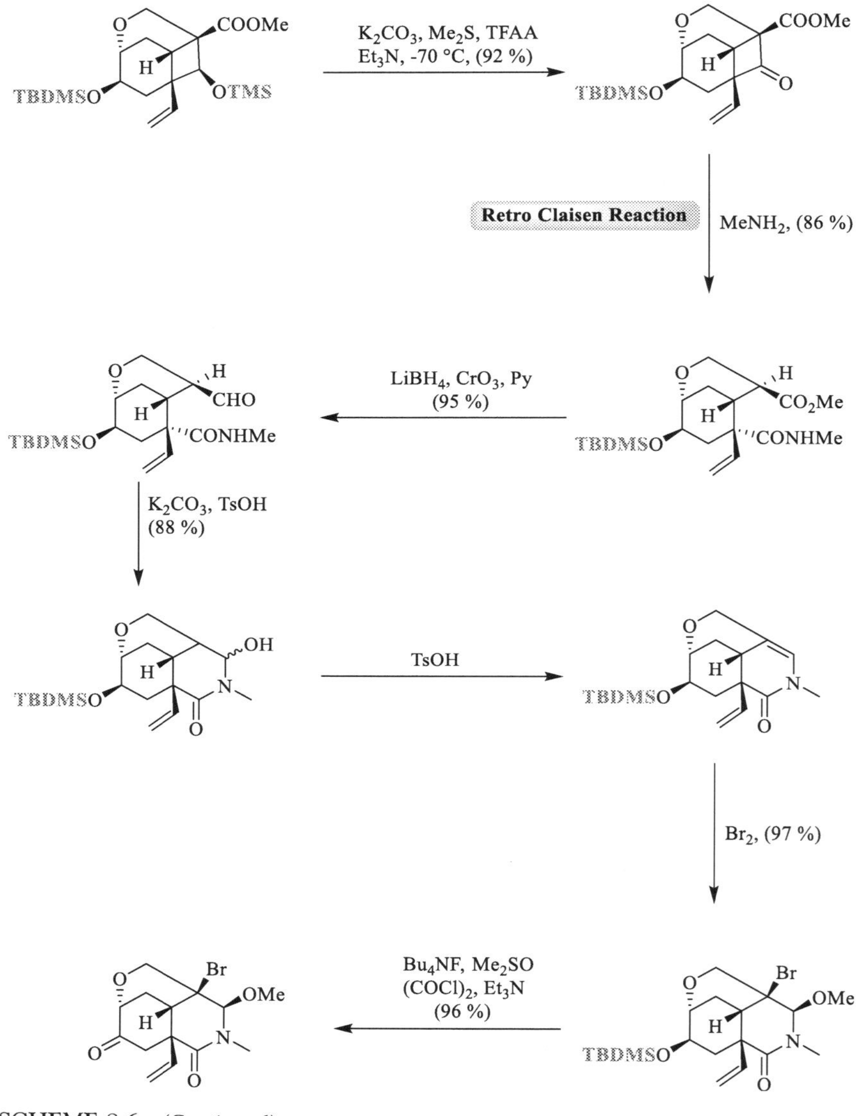

124

8. Gelsemine

SCHEME 8.6 (Continued).

Indole Alkaloids

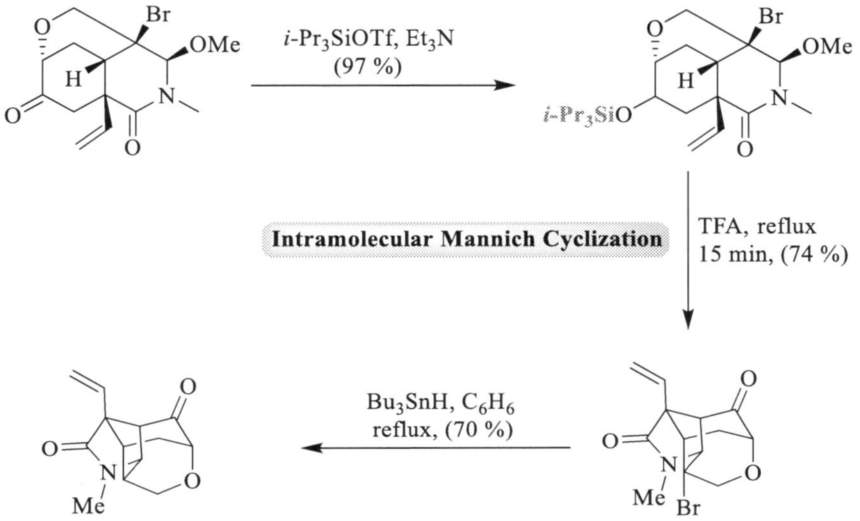

8.2 Formal syntheses of gelsemine

125

Intramolecular Mannich Cyclization

i-Pr$_3$SiOTf, Et$_3$N
(97 %)

TFA, reflux
15 min, (74 %)

Bu$_3$SnH, C$_6$H$_6$
reflux, (70 %)

SCHEME 8.6 (Continued).

Indole Alkaloids

126

8. Gelsemine

Key features

- Photoinduced intramolecular cycloaddition of triene to produce key tricyclic diester intermediate
- Reduction of tricyclic diester intermediate to corresponding diol
- Conversion of diol to acetoxy alcohol, corresponding acid and ester, respectively
- Synthesis of tricyclic selenide
- Oxidative deselenylation to produce alkene
- Desilylation leading to cyclobutanol
- Oxidation to cyclobutane-β-ketoester
- Retro Claisen reaction
- Chemoselective reduction and reoxidation
- Aldehyde epimerization
- Cyclization to produce hydroxyl lactam
- Dehydration to corresponding enamide
- Desilylation and oxidation to produce key intermediate keto lactam
- Conversion of keto lactam to enol silyl ether
- Conversion of enol silyl ether to bromoketone
- Intramolecular Mannich reaction
- Reductive debromination of the ketone to afford gelsemine intermediate

8.2.2 Johnson's spiroannelation of gelsemine (1994)

See Scheme 8.7.

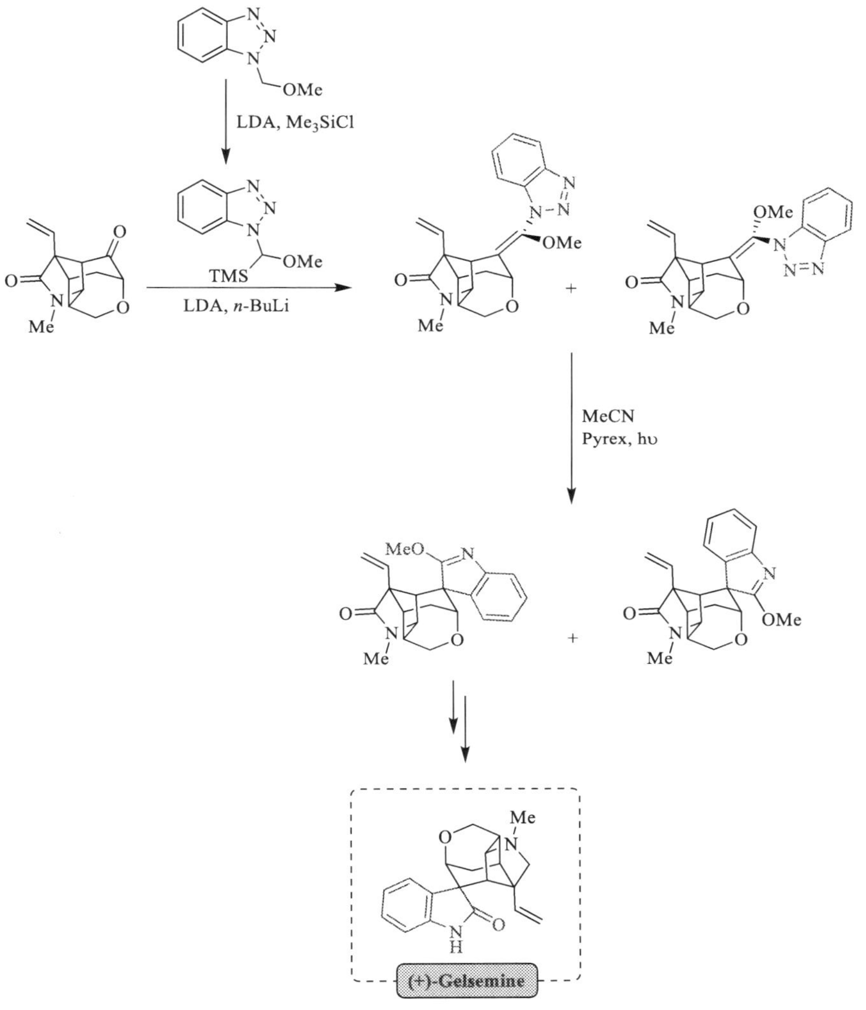

8.2 Formal syntheses of gelsemine

127

SCHEME 8.7 Johnson's spiroannelation of gelsemine.[14]

Key features

- Condensation of ketone intermediate of gelsemine with lithiated 1-(methoxytrimethylsily1methyl)benzotriazole to produce a mixture of (*E*)- and (*Z*)-methoxymethylene isomers
- Separate irradiation of both isomers in acetonitrile to produce gelsemine spiroannelated intermediates

128
8. Gelsemine

8.2.3 Zhou's synthesis of spirocyclopentaneoxindole intermediate through intramolecular Michael cyclization (2012)

See Scheme 8.8.

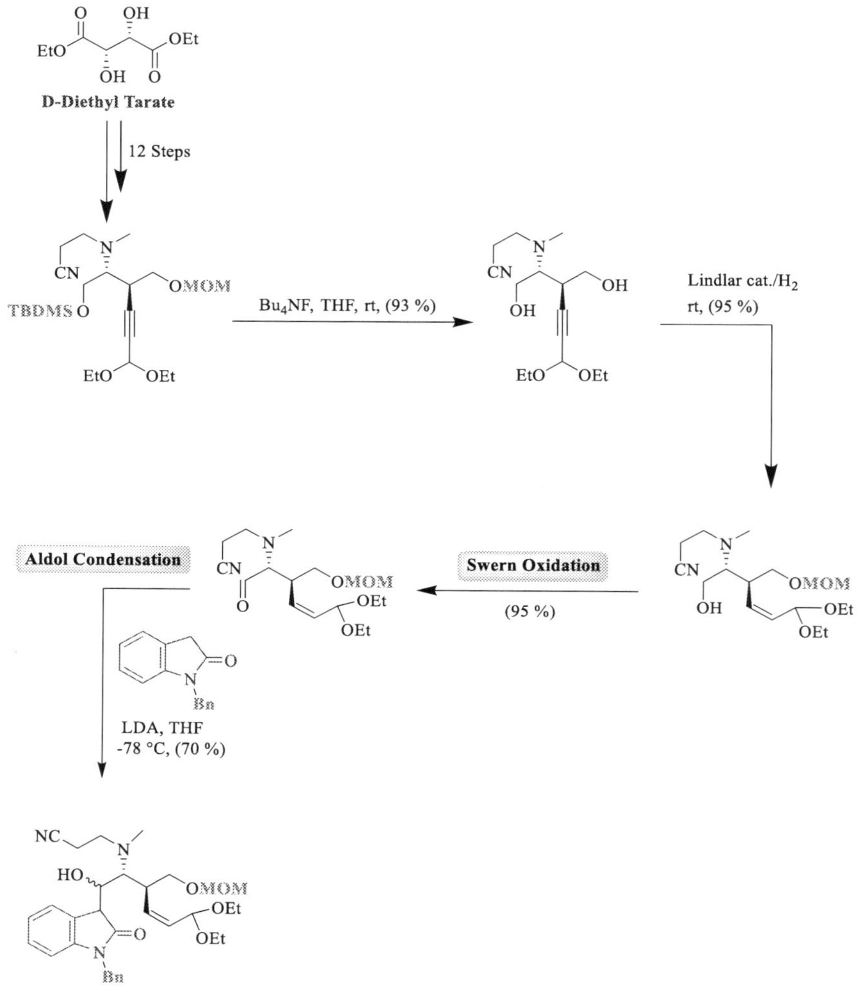

SCHEME 8.8 Synthesis of spirocyclopentaneoxindole intermediate for (+)-gelsemine.[15]

Indole Alkaloids

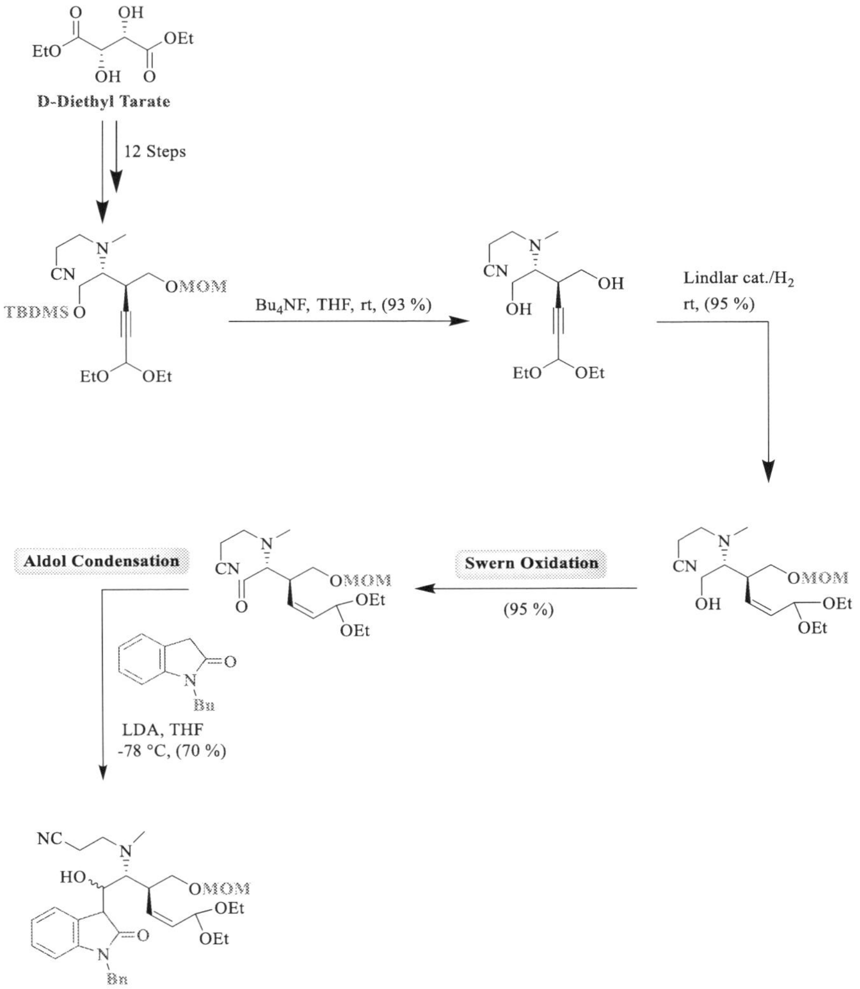

8.2 Formal syntheses of gelsemine · 129

SCHEME 8.8 (Continued).

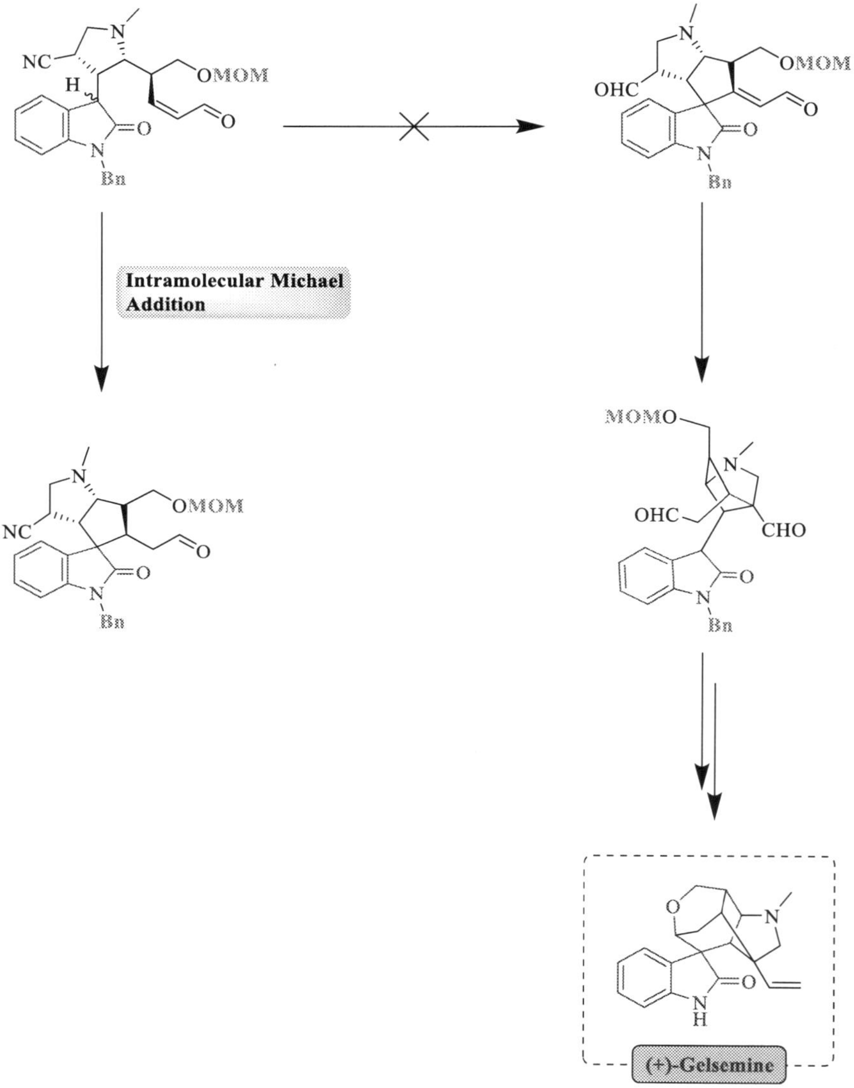

130

8. Gelsemine

SCHEME 8.8 (Continued).

Indole Alkaloids

Key features

- Two continuous intramolecular Michael additions to generate C and D rings and C20 quaternary stereocenter of (+)-gelsemine
- Produced intermediate failed to generate (+)-gelsemine
- Diethoxyprop-1-yn base protected starting material was produced from D-diethyl tarate in 12 steps
- Deprotection to produced alcohol followed by partial hydrogenation of triple bond to produce Z-conformation alkene
- Oxidation, aldol condensation with N-benzyloxindole, and subsequent dehydration
- Construction of pyrrolidine ring (D ring) via first intramolecular Michael addition of C20 to C6
- Synthesis of α,β-unsaturated aldehyde via acetal deprotection
- Intramolecular Michael addition of C7 to C15 to produce enantiomerically pure spirocyclopentaneoxindole derivative

8.2.4 Zhou's second attempt for spirocyclopentaneoxindole intermediate through intramolecular Michael cyclization (2012)

See Scheme 8.9.

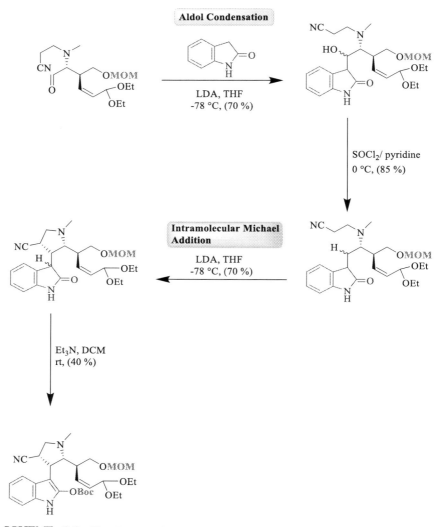

SCHEME 8.9 Zhou's second attempt for spirocyclopentaneoxindole intermediate through intramolecular Michael cyclization.[15]

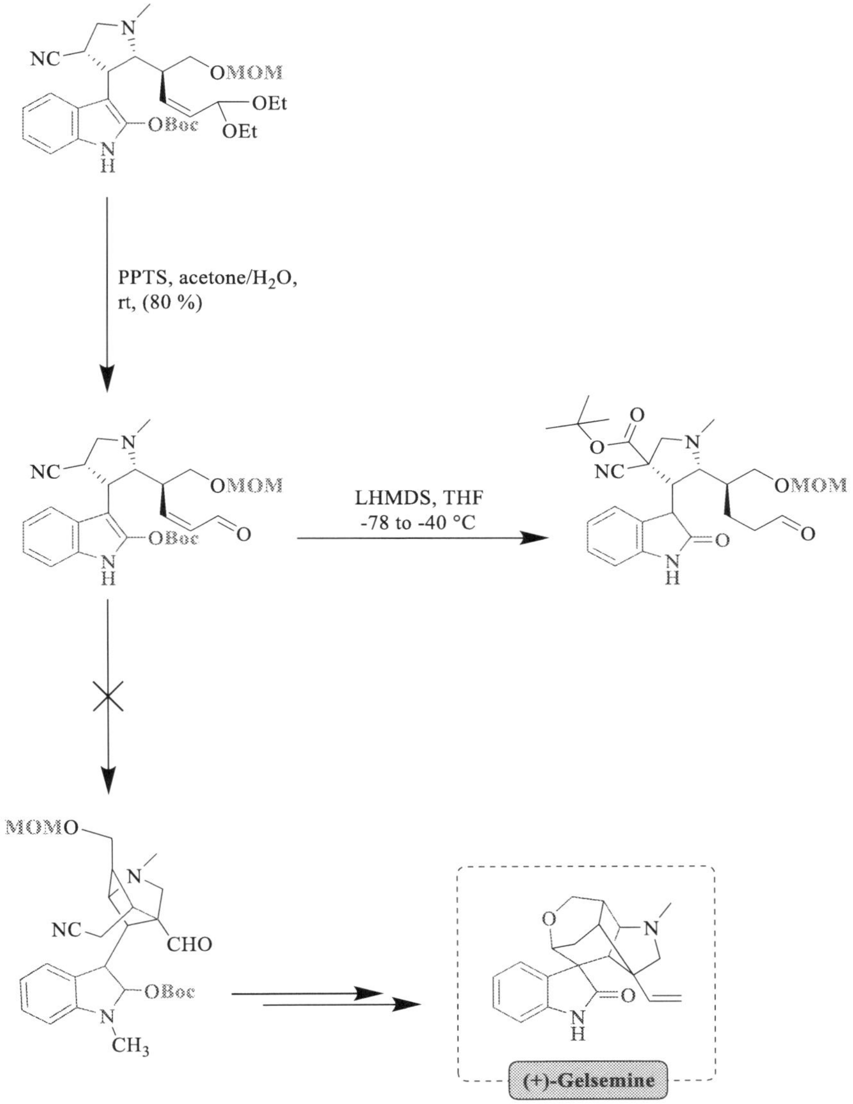

8.2 Formal syntheses of gelsemine

133

PPTS, acetone/H₂O,
rt, (80 %)

LHMDS, THF
-78 to -40 °C

SCHEME 8.9   (Continued).

Indole Alkaloids

134
8. Gelsemine

Key features

- Aldol condensation, Lindlar reduction, and Michael addition onto the aldehyde produced in the first scheme described in Section 8.2.3
- Protection of oxindole C = O functionality via ditertbutyl-dicarbonate
- Acetal removal to produce stable α,β-unsaturated aldehyde
- Intramolecular Michael addition between C20 and C15 using LHMDS to produce an enantiomerically pure spirocyclopentaneoxindole intermediate

References

1. Sonnenschein, F. L. Ueber Einige Bestandtheile von *Gelsemium sempervirens*. *Deut. Chem. Ges. Ber.* **1876,** *9* (2), 1182−1186.
2. Moore, C. W. CCXXXII.—The Constituents of *Gelsemium*. *J. Chem. Soc. Trans.* **1910,** *97,* 2223−2233.
3. Lovell, F.; Pepinsky, R.; Wilson, A. X-Ray Analysis of the Structure of Gelsemine Hydrohalides. *Tetrahedron Lett.* **1959,** *1* (4), 1−5.
4. Ponglux, D.; Wongseripipatana, S.; Subhadhirasakul, S.; Takayama, H.; Yokota, M.; Ogata, K.; Phisalaphong, C.; Aimi, N.; Sakai, S.-i Studies on the Indole Alkaloids of *Gelsem1um* elegans (Thailand): Structure Elucidation and Proposal of Biogenetic Route. *Tetrahedron* **1988,** *44* (16), 5075−5094.
5. Schun, Y.; Cordell, G. A. Rankinidine, A New Indole Alkaloid From *Gelsemium rankinii*. *J. Nat. Prod.* **1986,** *49* (5), 806−808.
6. Fukuyama, T.; Liu, G. Stereocontrolled Total Synthesis of (±)-Gelsemine. *J. Am. Chem. Soc.* **1996,** *118* (31), 7426−7427.
7. Fukuyama, T.; Liu, G. Stereocontrolled Total Synthesis of (±)-Gelsemine. *Pure Appl. Chem.* **1997,** *69* (3), 501−506.
8. Yokoshima, S.; Tokuyama, H.; Fukuyama, T. Enantioselective Total Synthesis of (+)-Gelsemine: Determination of Its Absolute Configuration. *Angew. Chem. Int. Ed.* **2000,** *112* (22), 4239−4241.
9. Madin, A.; O'Donnell, C. J.; Oh, T.; Old, D. W.; Overman, L. E.; Sharp, M. J. Total Synthesis of (±)-Gelsemine. *Angew. Chem. Int. Ed.* **1999,** *38* (19), 2934−2936.
10. Madin, A.; O'Donnell, C. J.; Oh, T.; Old, D. W.; Overman, L. E.; Sharp, M. J. Use of the Intramolecular Heck Reaction for Forming Congested Quaternary Carbon Stereocenters. Stereocontrolled Total Synthesis of (±)-Gelsemine. *J. Am. Chem. Soc.* **2005,** *127* (51), 18054−18065.
11. Zhou, X.; Xiao, T.; Iwama, Y.; Qin, Y. Biomimetic Total Synthesis of (+)-Gelsemine. *Angew. Chem. Int. Ed.* **2012,** *51* (20), 4909−4912.
12. Chen, X.; Duan, S.; Tao, C.; Zhai, H.; Qiu, F. G. Total Synthesis of (+)-Gelsemine via an Organocatalytic Diels−Alder Approach. *Nat. Commun.* **2015,** *6* (1), 1−7.
13. Sheikh, Z.; Steel, R.; Tasker, A. S.; Johnson, A. P. A Total Synthesis of Gelsemine: Synthesis of a Key Tetracyclic Intermediate. *J. Chem. Soc. Chem. Commun.* **1994,** (6), 763−764.
14. Dutton, J. K.; Steel, R. W.; Tasker, A. S.; Popsavin, V.; Johnson, A. P. A Total Synthesis of Gelsemine: Oxindole Spiroannelation. *J. Chem. Soc. Chem. Commun.* **1994,** (6), 765−766.
15. Zhou, S.; Xiao, T.; Song, H.; Zhou, X. Studies Toward the Total Synthesis of (+)-Gelsemine and Synthesis of Spirocyclopentaneoxindole Through Intramolecular Michael Cyclization. *Tetrahedron Lett.* **2012,** *53* (42), 5684−5687.

CHAPTER 9

Paraherquamide A and B

The paraherquamides (Fig. 9.1) belong to an infrequent subclass of fungus-derived natural products containing a bicyclo[2.2.2]diazaoctane framework substituted with proline moiety. Paraherquamide A was first isolated by Yamazaki and coworkers in 1981 from cultures of *Penicillium paraherquei*.[1] Later on, paraherquamides B—G have been isolated from various *Aspergillus* and *Penicillium* species.

These molecules, like other spirooxindoles, have attracted substantial attention due to their molecular intricacy, intriguing biogenesis, and significant biological activities. Paraherquamide A exhibits potent anthelmintic and antinematodal activities.[2] These molecules also represent a totally new structural class of anthelmintic compounds. In addition, they serve as potent drugs for the treatment of intestinal parasites in animals.[3] Their mode of action is incompletely characterized till yet, but the latest research proposes that they are selective reasonable cholinergic antagonists.[4]

Inspiring by this vast biological spectrum, many efforts have been reported for in lab synthesis of this class. Up to now, two total synthetic approaches, one formal and one semisynthesis starting from marcfortine, are reported, which are described in detail in this section.

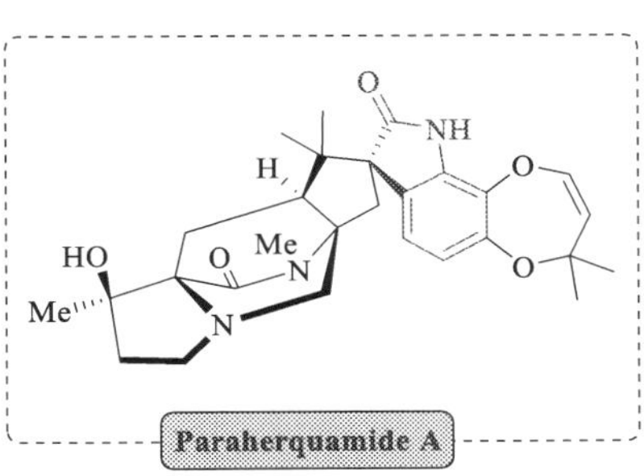

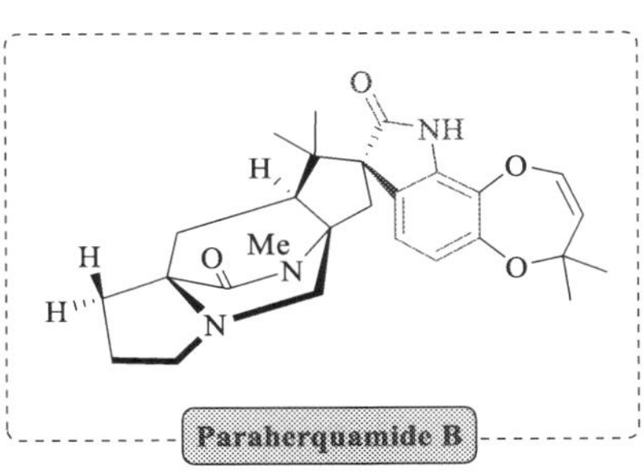

FIGURE 9.1   Structures of paraherquamides A and B.

Indole Alkaloids
DOI: https://doi.org/10.1016/B978-0-323-91674-5.00003-1

9. Paraherquamide A and B

9.1 Total syntheses of paraherquamides A and B

9.1.1 Williams' asymmetric stereocontrolled first total synthesis of paraherquamide A (1996, 2000, 2003)

See Scheme 9.1.

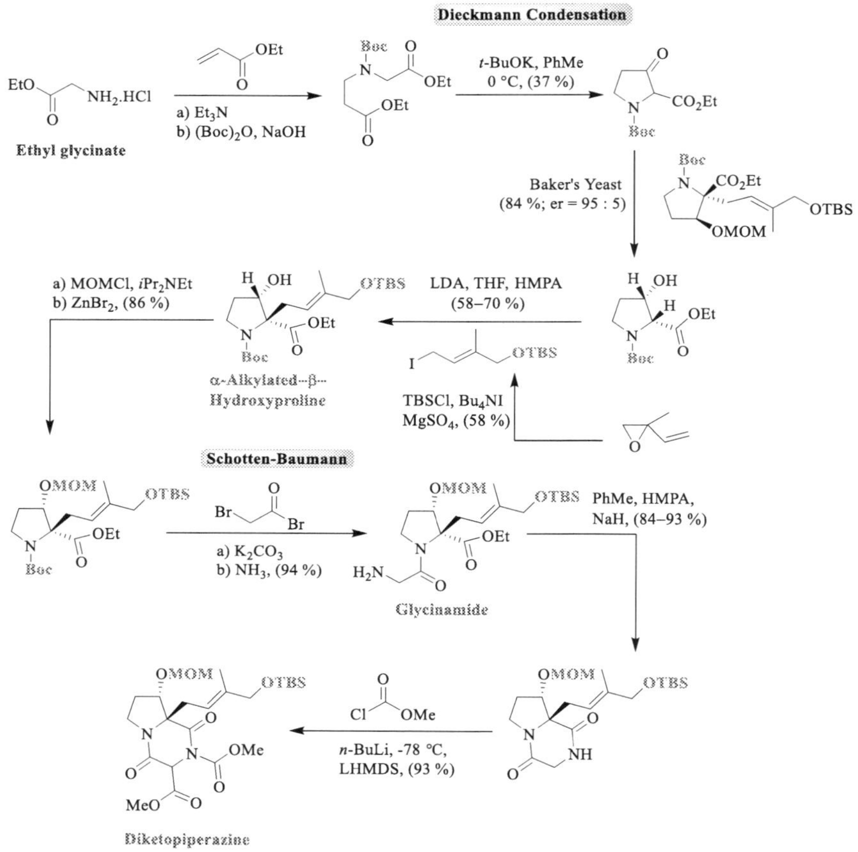

SCHEME 9.1 Williams' asymmetric stereocontrolled first total synthesis of paraherquamide A.[5–7]

Indole Alkaloids

9.1 Total syntheses of paraherquamides A and B

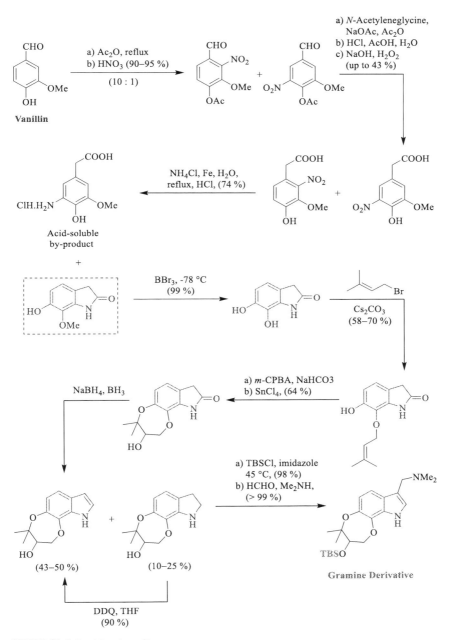

SCHEME 9.1 (Continued).

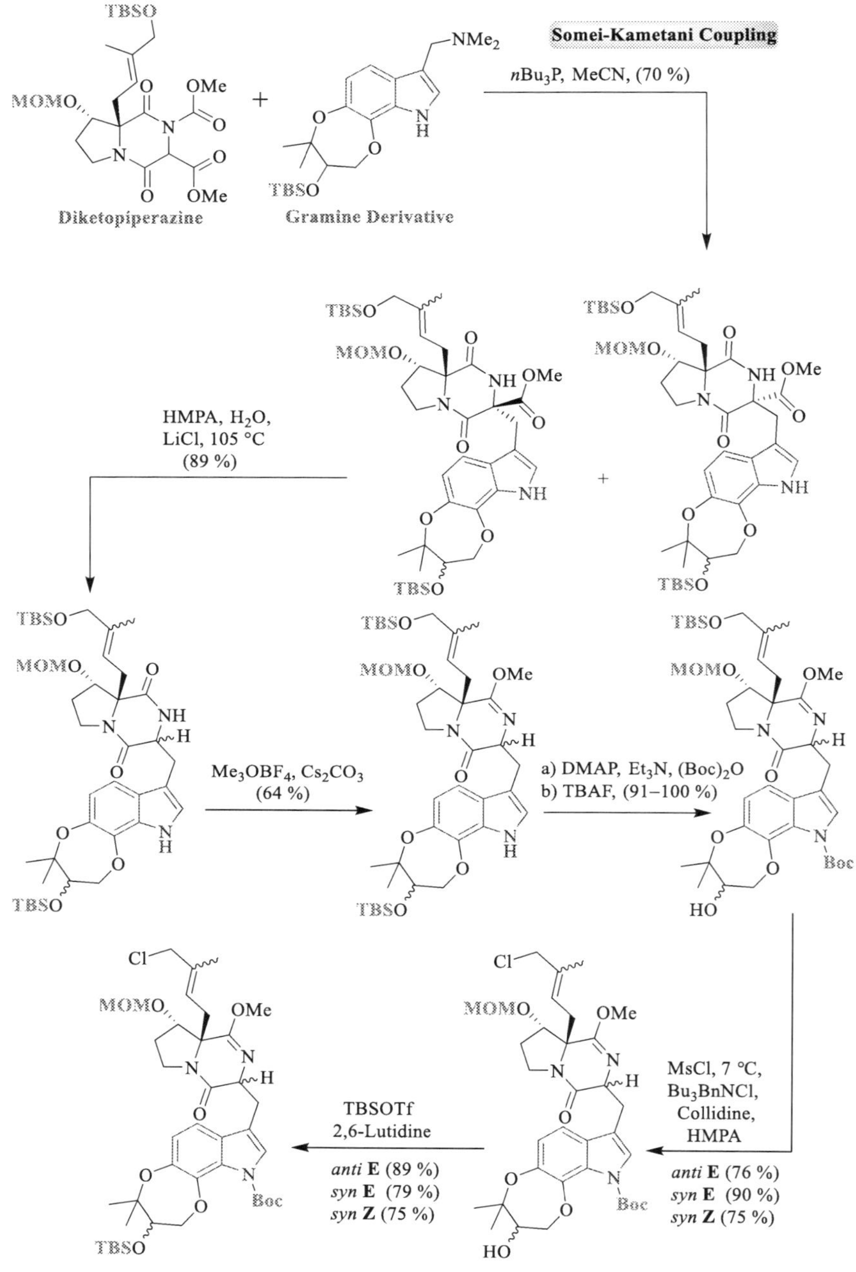

138
9. Paraherquamide A and B

SCHEME 9.1 (Continued).

Indole Alkaloids

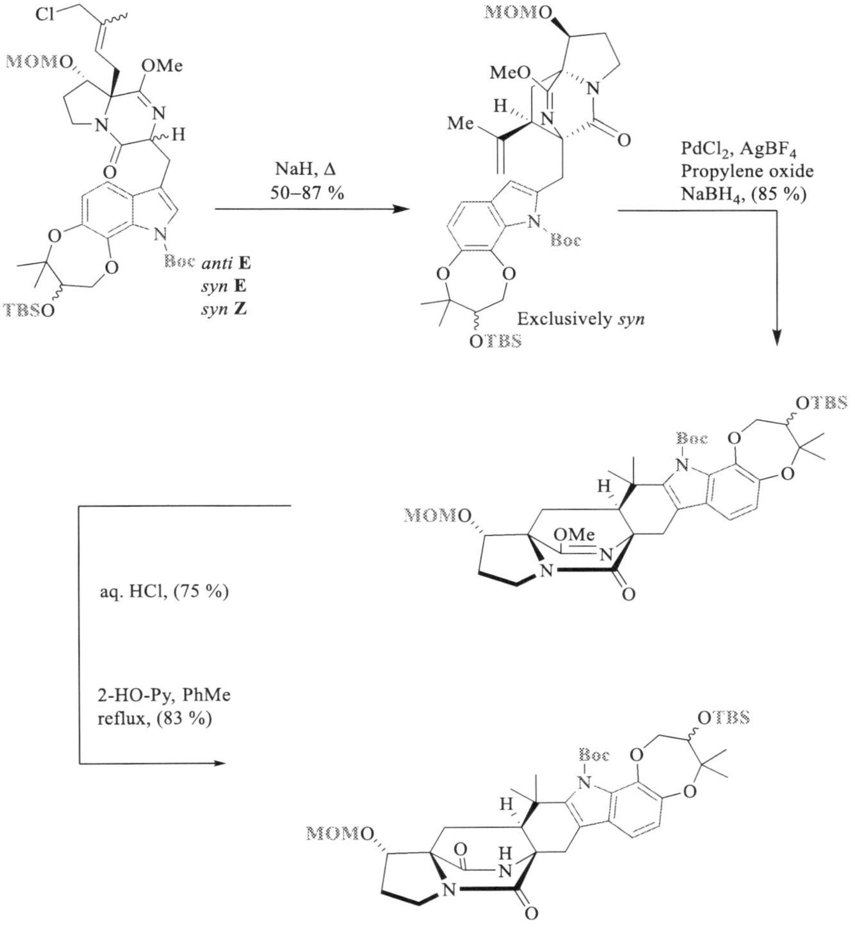

9.1 Total syntheses of paraherquamides A and B 139

SCHEME 9.1 (Continued).

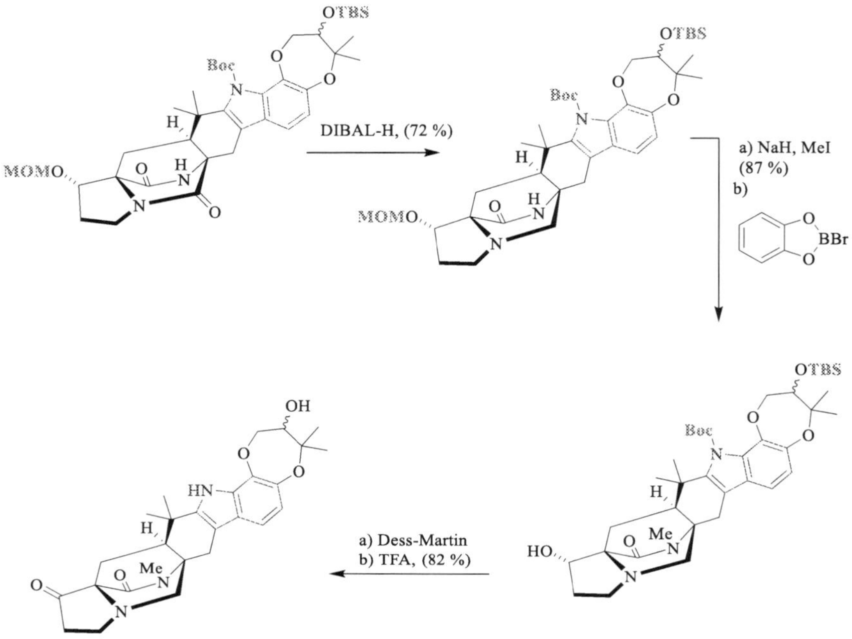

9. Paraherquamide A and B

SCHEME 9.1 (Continued).

Indole Alkaloids

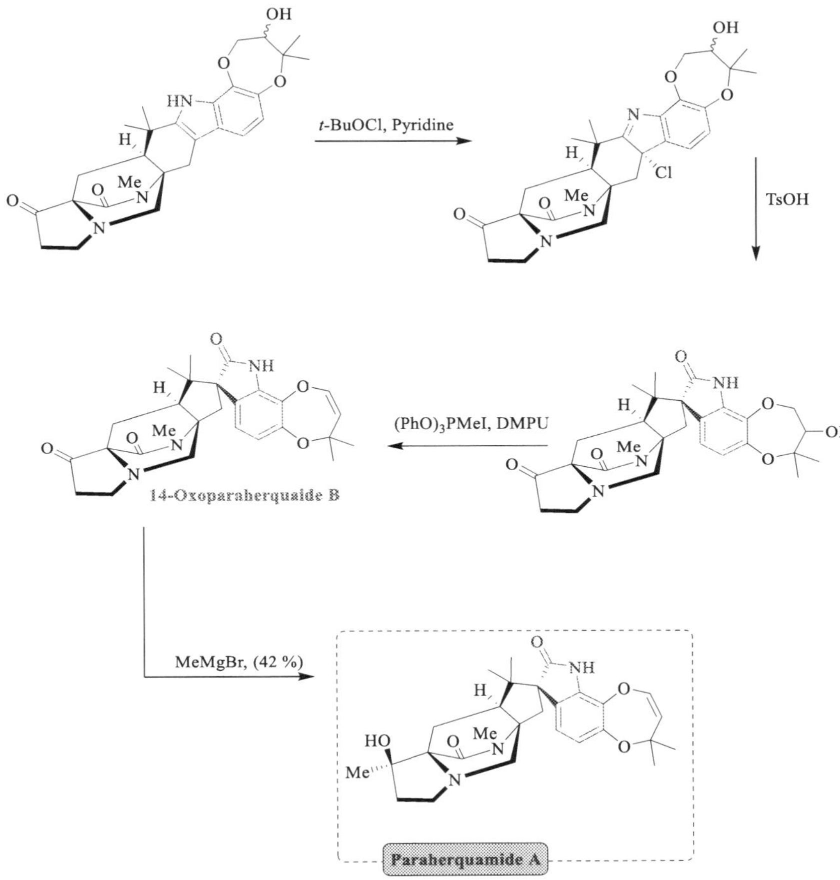

9.1 Total syntheses of paraherquamides A and B

SCHEME 9.1 (Continued).

Key features

- Enantioselective synthesis of an α-alkylated-β-hydroxyproline
- Synthesis of diketopiperazine
- Synthesis of gramine (indole) derivative
- Somei–Kametani coupling of indole and diketopiperazine
- Formation of the heptacycle
- Spirocyclization

Indole Alkaloids

142

9.1.2 Williams and Cushing convergent, stereocontrolled, asymmetric total synthesis of (+)- paraherquamide B (1996)

See Scheme 9.2.

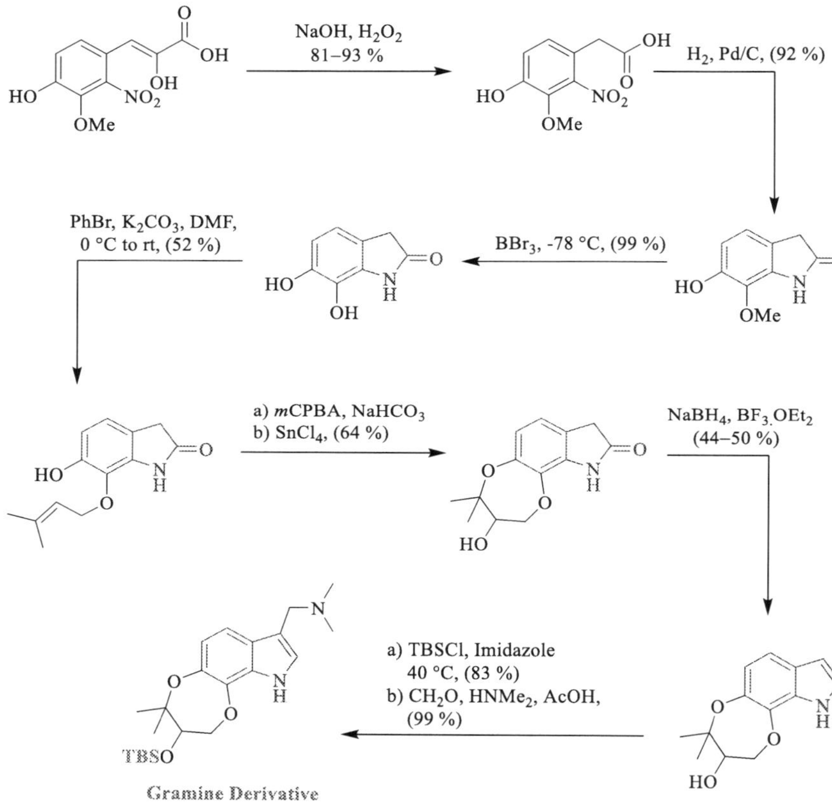

SCHEME 9.2 Williams and Cushing convergent, stereocontrolled, asymmetric total synthesis of paraherquamide B.[8]

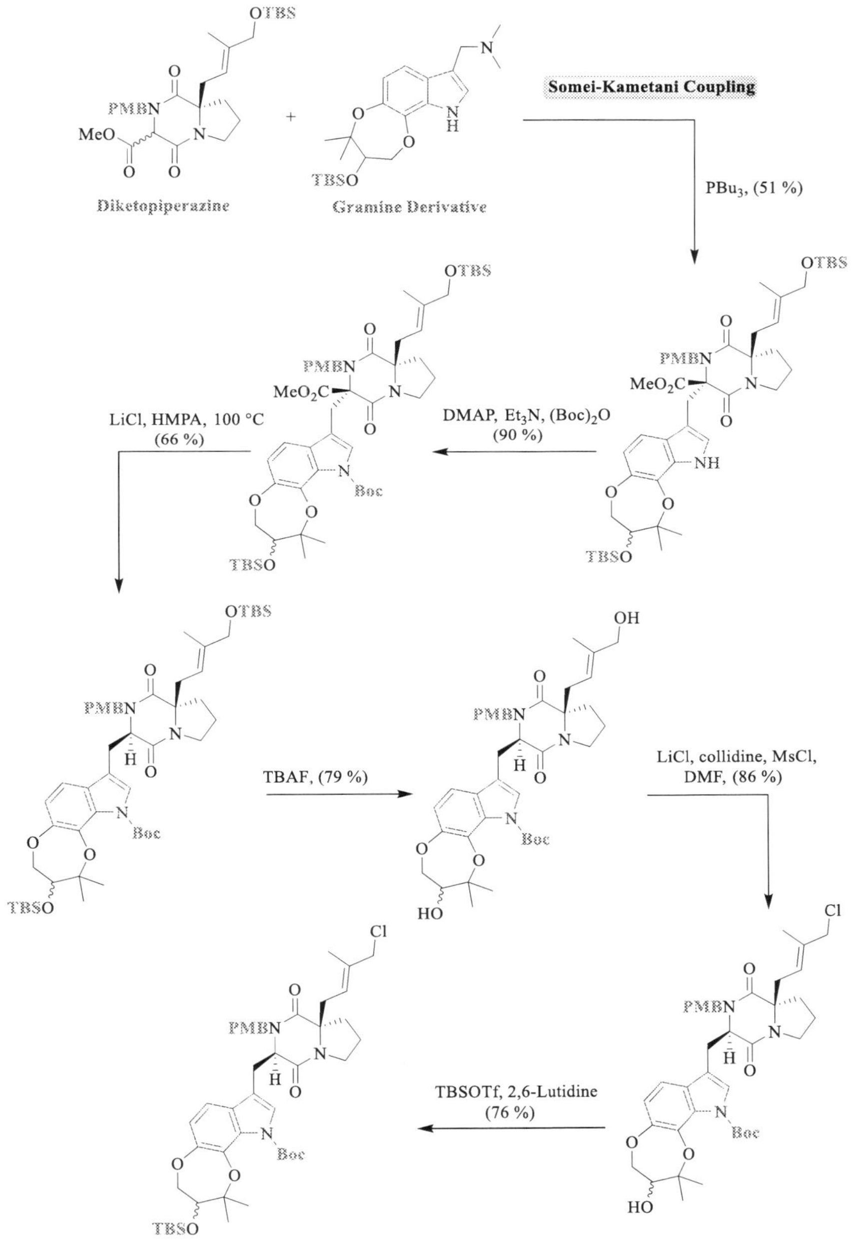

9.1 Total syntheses of paraherquamides A and B 143

SCHEME 9.2 (Continued).

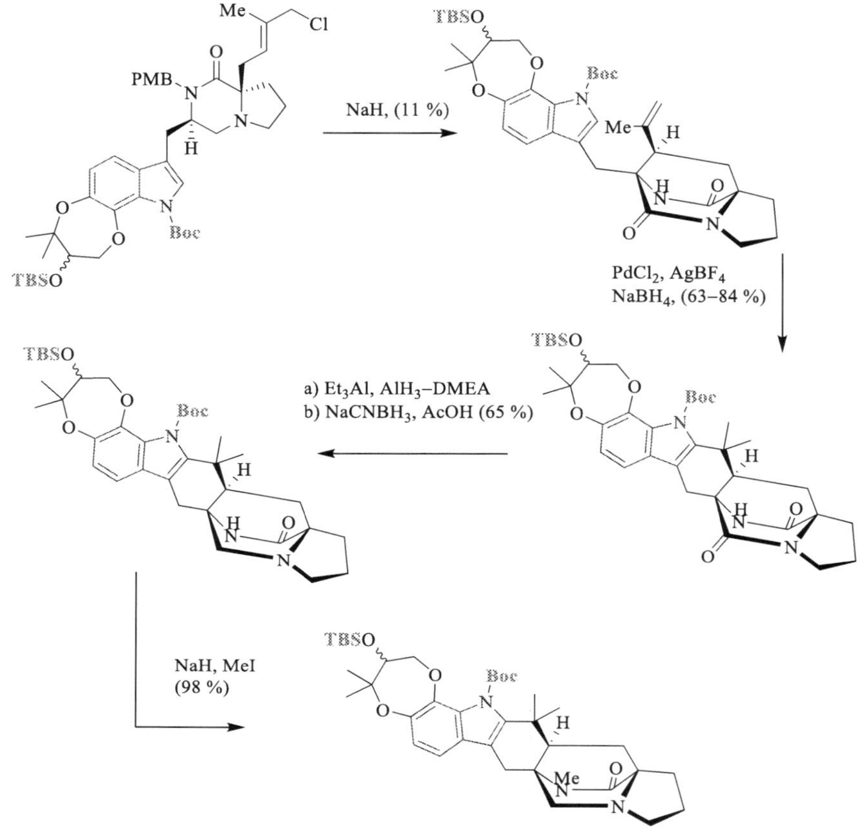

9. Paraherquamide A and B

SCHEME 9.2 (Continued).

Indole Alkaloids

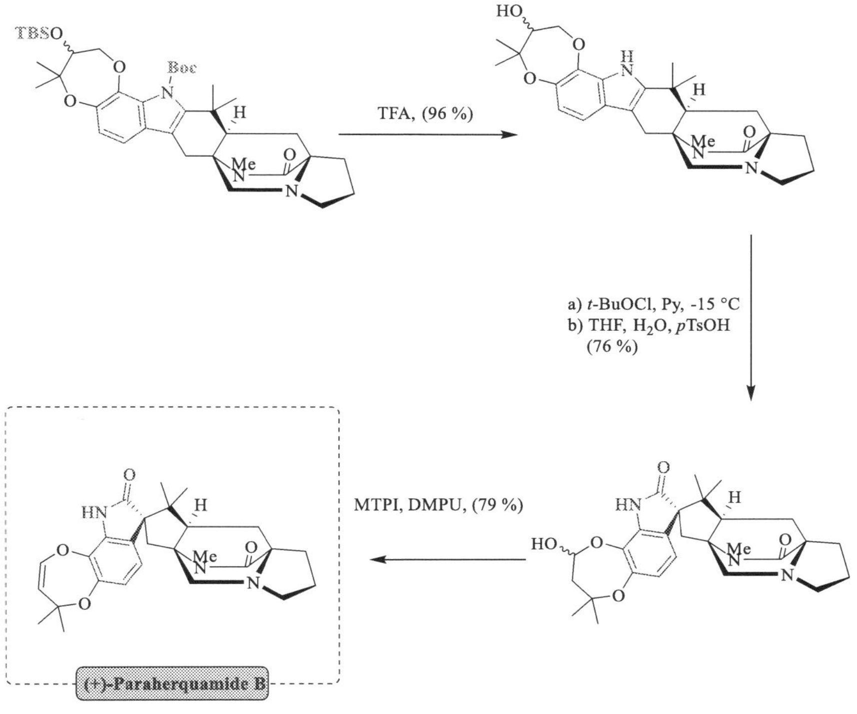

9.2 McWhorter's formal synthesis

SCHEME 9.2 (Continued).

Key features

- An improved procedure for the reduction of unprotected oxindoles to indoles
- Somei—Kametani coupling reaction of indole and diketopiperazine
- Stereocontrolled intramolecular S_N2' cyclization reaction to construct the core bicyclo[2.2.2] ring system
- A mild Pd(II)-mediated cyclization reaction to furnish complex tetrahydrocarbazole
- Chemoselective reduction of highly hindered tertiary lactam

9.2 McWhorter's formal synthesis of 6,7-dihydroxyoxindole; a subunit of paraherquamide A (1996)

See Scheme 9.3.

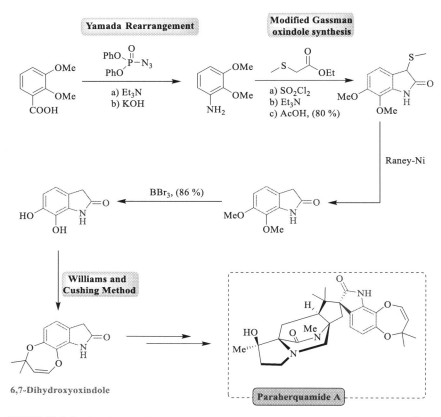

SCHEME 9.3 Synthesis of 6,7-dihydroxyoxindole; a subunit of paraherquamide A.[9]

Key features

- Synthesis of 6,7-dihydroxyoxindole in four steps with an overall yield of 35% from 2,3-dimethoxybenzoic acid
- Conversion of 2,3-dimethoxybenzoic acid to 2,3-dimethoxyaniline by Yamada modification of the Curtius rearrangement
- Modified Gassman oxindole synthesis
- Desulfurization of synthesized oxindole and subsequent demethylation to furnish 6,7-dihydroxyoxindole

9.3 Conversion of marcfortine A to paraherquamide A via paraherquamide B (1997)

See Scheme 9.4.

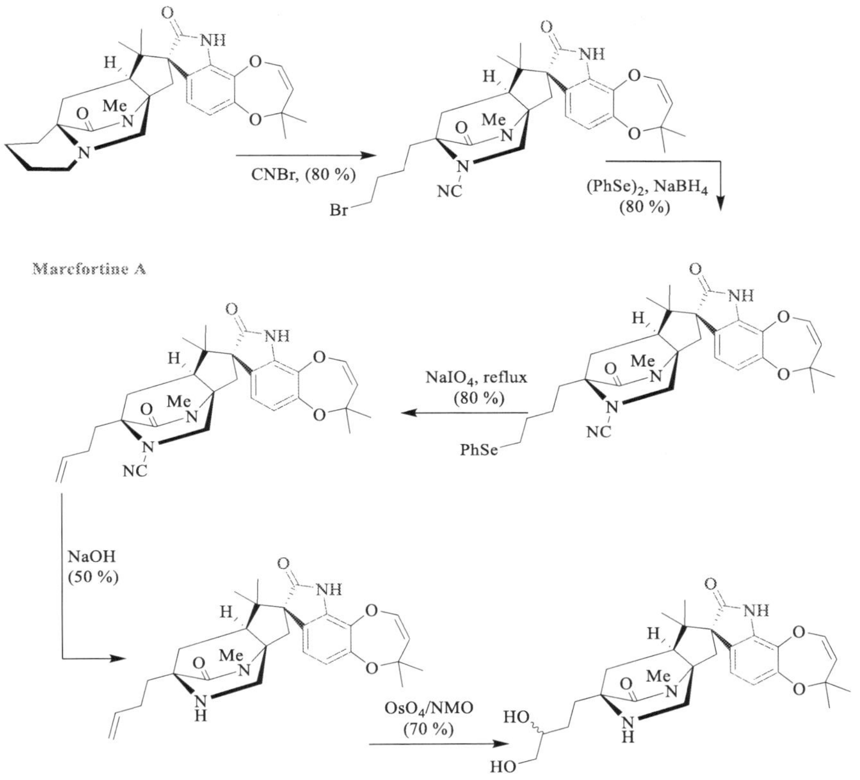

9.3 Conversion of marcfortine A to paraherquamide A via paraherquamide B (1997)   147

SCHEME 9.4   Conversion of marcfortine A to paraherquamide A via paraherquamide B.[10]

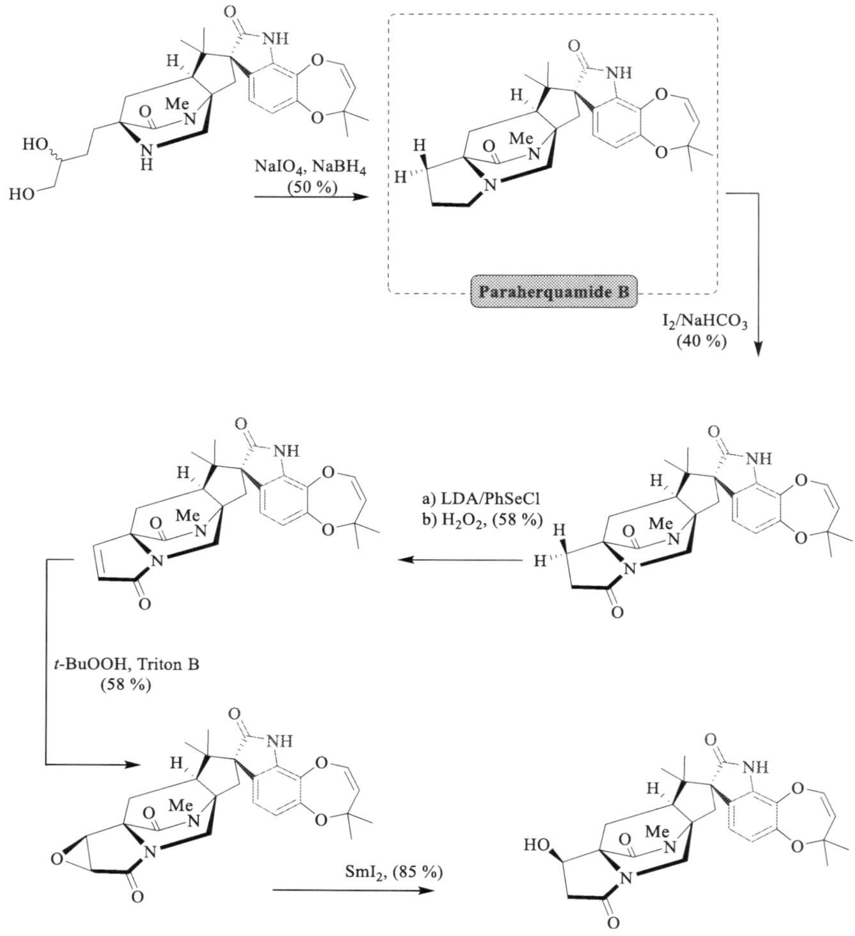

9. Paraherquamide A and B

SCHEME 9.4   (Continued).

Indole Alkaloids

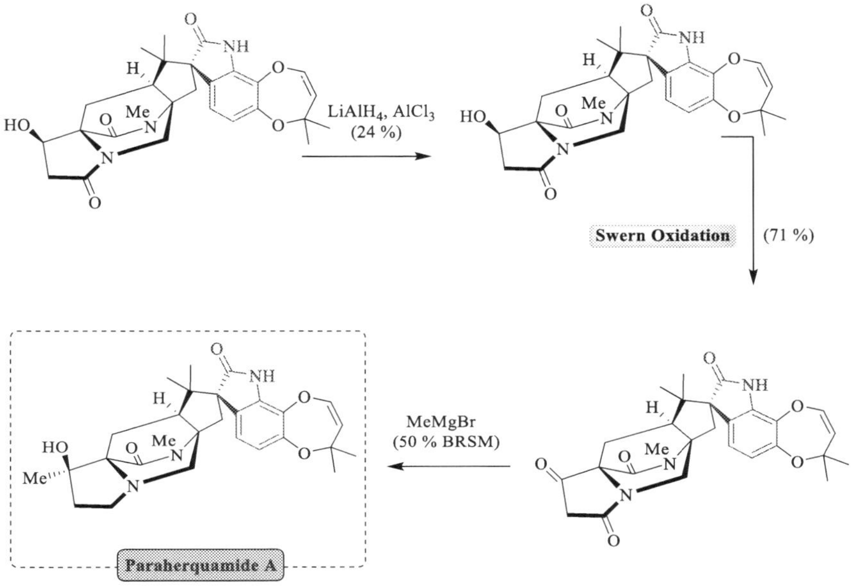

SCHEME 9.4 (Continued).

Key features

- Opening the G-ring of marcfortine A, oxidatively removing one carbon atom, and reclosing the ring to furnish paraherquamide B in seven steps
- Conversion of paraherquamide B to paraherquamide A in six steps resulting 50% yield (BRSM)

References

1. Yamazaki, M.; Okuyama, E.; Kobayashi, M.; Inoue, H. The Structure of Paraherquamide, A Toxic Metabolite From *Penicillium paraherquei*. *Tetrahedron Lett.* **1981,** *22* (2), 135–136.
2. Shoop, W.; Eary, C.; Michael, B.; Haines, H.; Seward, R. Anthelmintic Activity of Paraherquamide in Dogs. *Vet. Parasitol.* **1991,** *40* (3–4), 339–341.
3. Blizzard, T. A.; Margiatto, G.; Mrozik, H.; Schaeffer, J. M.; Fisher, M. H. Chemical Modification of Paraherquamide. 3. Vinyl Ether Modified Analogs. *Tetrahedron Lett.* **1991,** *32* (22), 2437–2440.
4. Zinser, E. W.; Wolf, M. L.; Alexander-Bowman, S. J.; Thomas, E. M.; Davis, J. P.; Groppi, V. E., et al. Anthelmintic Paraherquamides Are Cholinergic Antagonists in Gastrointestinal Nematodes and Mammals. *J. Vet. Pharmacol. Ther.* **2002,** *25* (4), 241–250.

150

5. Williams, R. M.; Cao, J. Studies on the Total Synthesis of Paraherquamide A. Stereocontrolled, Asymmetric Synthesis of α-Alkyl-3-hydroxyproline Derivatives. *Tetrahedron Lett.* **1996,** *37* (31), 5441−5444.
6. Williams, R. M.; Cao, J.; Tsujishima, H. Asymmetric, Stereocontrolled Total Synthesis of Paraherquamide A. *Angew. Chem. Int. Ed.* **2000,** *39* (14), 2540−2544.
7. Williams, R. M.; Cao, J.; Tsujishima, H.; Cox, R. J. Asymmetric, Stereocontrolled Total Synthesis of Paraherquamide A. *J. Am. Chem. Soc.* **2003,** *125* (40), 12172−12178.
8. Cushing, T. D.; Sanz-Cervera, J. F.; Williams, R. M. Stereocontrolled Total Synthesis of (+)-Paraherquamide B. *J. Am. Chem. Soc.* **1996,** *118* (3), 557−579.
9. Savall, B. M.; McWhorter, W. W. Synthesis of 6, 7-Dihydroxyoxindole (A Subunit of Paraherquamide A). *J. Org. Chem.* **1996,** *61* (24), 8696−8697.
10. Lee, B. H.; Clothier, M. F. Conversion of Marcfortine A to Paraherquamide A via Paraherquamide B. The First Formal Synthesis of Paraherquamide A. *J. Org. Chem.* **1997,** *62* (6), 1795−1798.

Indole Alkaloids

CHAPTER 10

Rhynchophylline and isorhynchophylline

Rhynchophylline and isorhynchophylline (Fig. 10.1) were isolated from *Uncaria rhynchophylla*. These are active alkaloids used widely in traditional Chinese medicine to cure hypertension and stroke.

A large number of biological properties are associated with these special tetracyclic spirooxindoles like hypotensive and antihypertensive activities, effective Ca^{2+} channel antagonist,[1] anticoagulants, neuroprotective activities,[2] vascular smooth muscle cell proliferation,[3] and receptor tyrosine kinase EphA4 inhibitors. In addition, these are also used to rescue hippocampal synaptic dysfunctions in AD.[4]

Due to vast biological properties, synthetic chemists are showing growing interest toward the syntheses of these molecules in lab. Up to now two racemic total syntheses, five formal syntheses, including two racemic and three enantioselective ones, and two semisyntheses have been described.

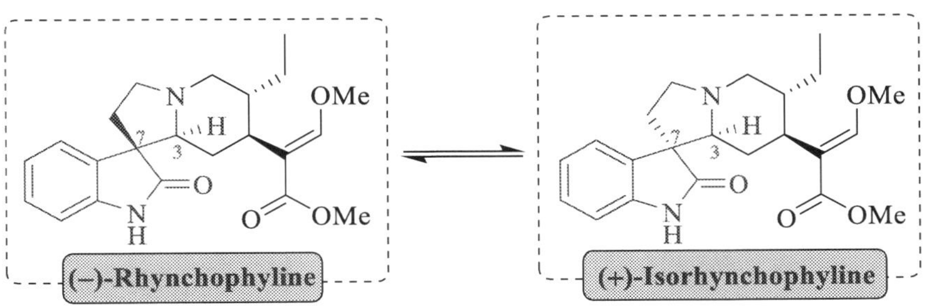

FIGURE 10.1   Structures of rhynchophylline and isorhynchophylline.

Indole Alkaloids
DOI: https://doi.org/10.1016/B978-0-323-91674-5.00008-0

152
10. Rhynchophylline and isorhynchophylline

10.1 Total syntheses of (−)-rhynchophylline and (+)-isorhynchophylline

10.1.1 Ban's first total synthesis (1972, 1975)

See Scheme 10.1.

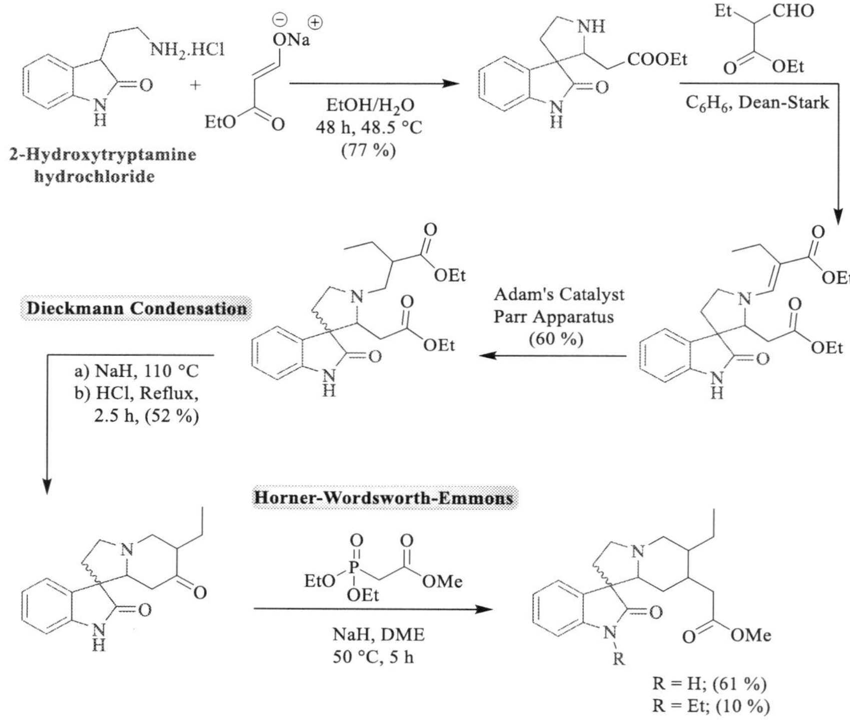

SCHEME 10.1 Ban's first total synthesis of (±)-rhynchophylline and (±)-isorhynchophylline.[5,6]

Indole Alkaloids

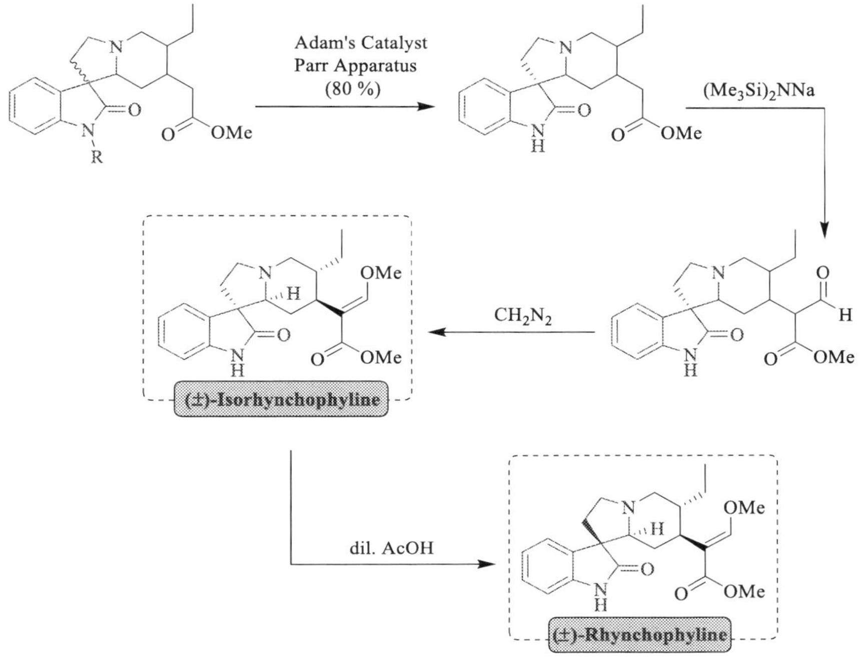

10.1 Total syntheses of (−)-rhynchophylline and (+)-isorhynchophylline

SCHEME 10.1 (Continued).

Key features

- Spirocyclization by condensation on 2-hydroxytryptamine hydrochloride
- Condensation of resultant spiro-compound with α-formylbutyrate and subsequent reduction of resultant condensed compound
- Dieckmann Condensation
- Horner−Wordsworth−Emmons reaction
- Formylation and subsequent methylation of formyl group to yield (+)-isorhychophylline
- Isomerization of (+)-isorhychophylline to (−)-rhynchophylline

10.1.2 Hiemstra's total synthesis (2013)

See Scheme 10.2.

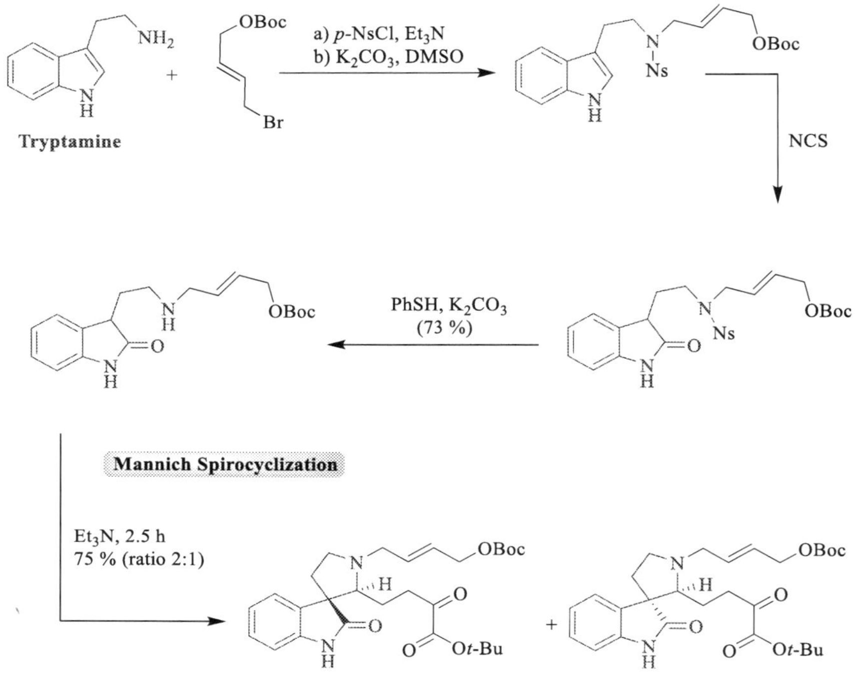

SCHEME 10.2 Hiemstra's total synthesis of (−)-rhynchophylline and (+)-isorhynchophylline.[7]

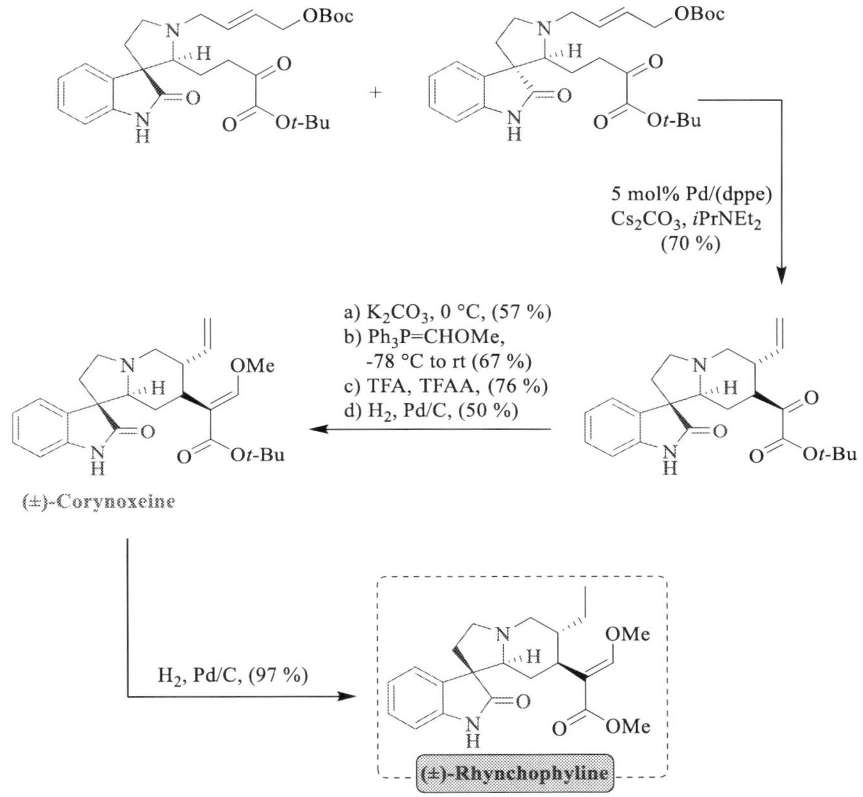

10.1 Total syntheses of (−)-rhynchophylline and (+)-isorhynchophylline

SCHEME 10.2 (Continued).

Key features

- Butenylation of tryptamine
- Mannich spirocyclization
- Cyclization of an α-keto ester enolate onto an allylic carbonate
- Synthesis of (±)-corynoxeine
- Conversion of (±)-corynoxeine to (±)-rhynchophylline via hydrogenation

10.1.3 Tong's total synthesis via Carreira's diastereoselective [3 + 2] annulation of cyclopropyl oxindole and highly functionalized cyclic aldimine (2019)

See Scheme 10.3.

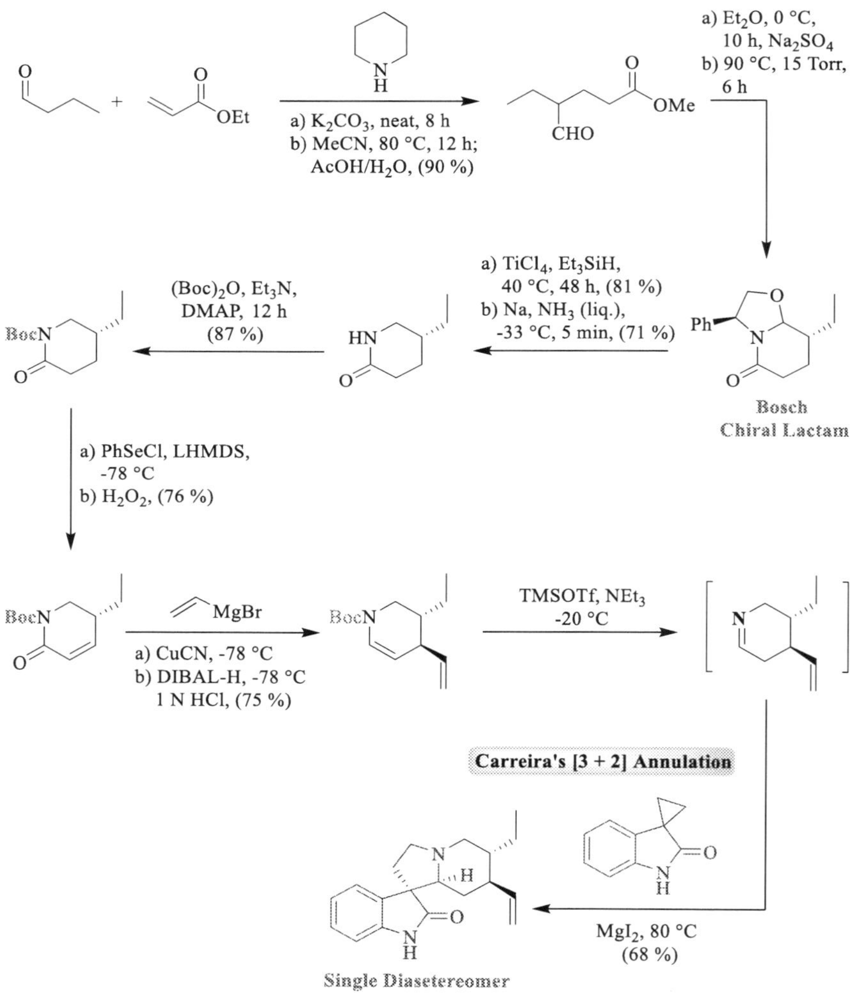

SCHEME 10.3 Tong's formal synthesis of (−)-rhynchophylline and (+)-isorhynchophylline.[8]

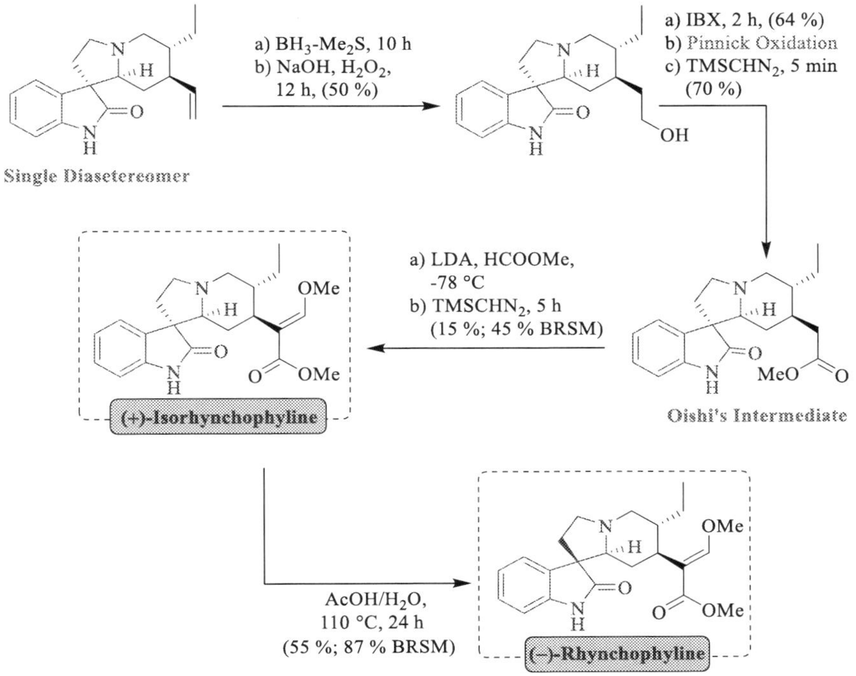

10.1 Total syntheses of (−)-rhynchophylline and (+)-isorhynchophylline

157

SCHEME 10.3 (Continued).

Key features

- Michael addition of aldehyde and α,β-unsaturated ester
- Lactamization to yield Bosch chiral lactam
- Reductive removal of chiral auxiliary
- Protection of δ-lactam followed by subsequtent sharpless α,β-unsaturation to lactam
- Michael addition of allyl Grignard reagent to lactam
- Carreira's diastereoselective [3 + 2] annulation
- Hydroboration/oxidation of the terminal alkene and resultant alcohol
- Pinnick oxidation
- Esterification of carboxylic acid to yield Ban's Intermediate
- Conversion of Ban's intermediate to (+)-isorhychophylline and its isomerization to (−)-rhynchophylline

Indole Alkaloids

10.1.3.1 Tong's enantioselective total synthesis of isorhynchophylline (2019)

See Scheme 10.4.

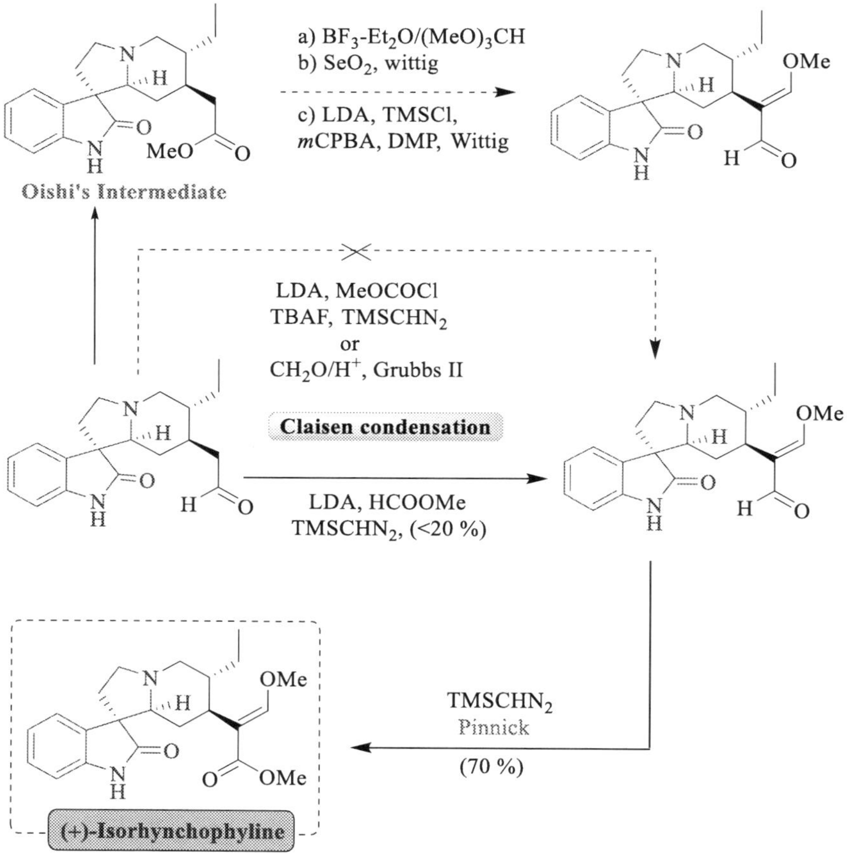

SCHEME 10.4 Tong's enantioselective total synthesis of isorhynchophylline.[8]

Key features

- Claisen condensation of spirooxindole intermediate with methyl formate
- Pinnick oxidation
- Esterification of side chain

10.2 Semisyntheses of (−)-rhynchophylline and (+)-isorhynchophylline

10.2.1 Finch's semisynthesis via oxidative transformations of indole alkaloid (1962)

See Scheme 10.5.

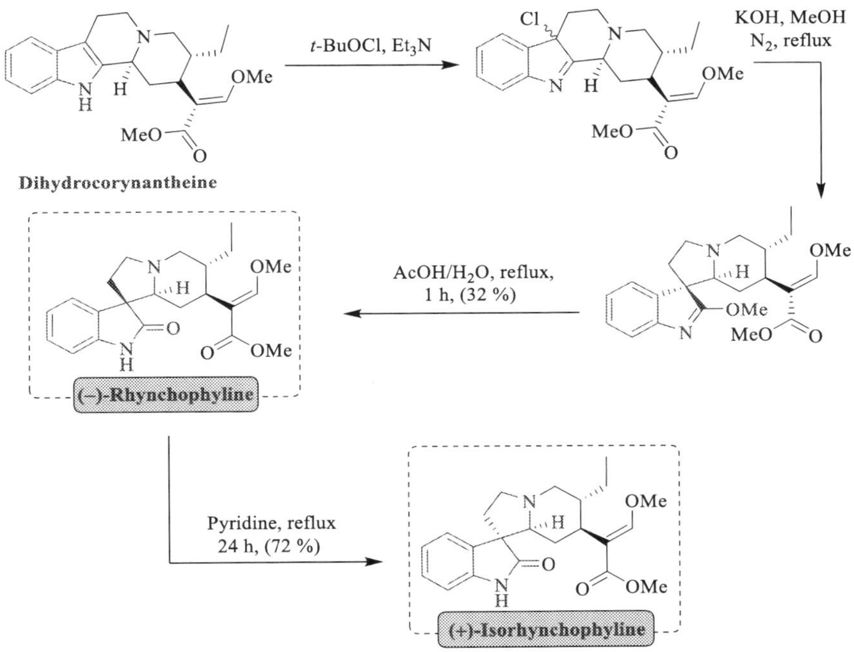

SCHEME 10.5 Finch's semisynthesis of (−)-rhynchophylline and (+)-isorhynchophylline via oxidative transformations of indole alkaloid.[9]

Key features

- Chlorination of dihydrocorynantheine; an indole alkaloid
- Methanolysis of resultant chloro-derivative
- Oxidation of methylated spirooxindole
- Isomerization of (−)-rhynchophylline to (+)-isorhychophylline

10.2.2 Brown's semisynthesis from dihydrosecologanin aglycone (1976)

See Scheme 10.6.

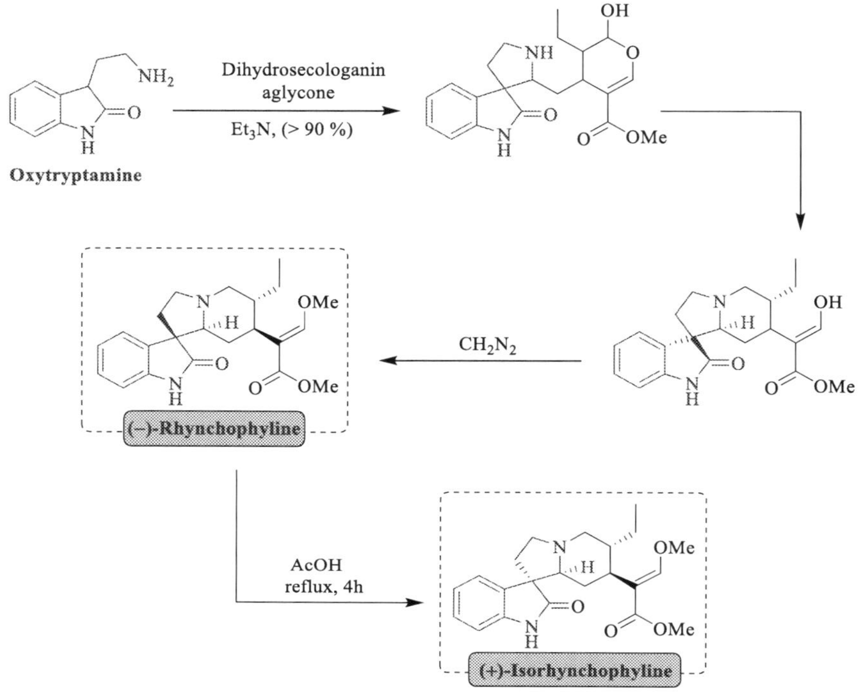

10. Rhynchophylline and isorhynchophylline

SCHEME 10.6 Brown's semisynthesis of (−)-rhynchophylline and (+)-isorhynchophylline starting from dihydrosecologanin aglycone.[10]

Key features

- Spirocyclization on oxytryptamine by addition of dihydrosecologanin aglycone
- Catalytic hydrogenation of oxydihydromancunine
- Methylation of formyl group to yield (+)-isorhychophylline
- Isomerization of (+)-isorhychophylline to (−)-rhynchophylline

10.3 Formal syntheses of (−)-rhynchophylline and (+)-isorhynchophylline

10.3.1 Martin's racemic formal synthesis via two ring-closing metathesis reactions (2006)

See Scheme 10.7.

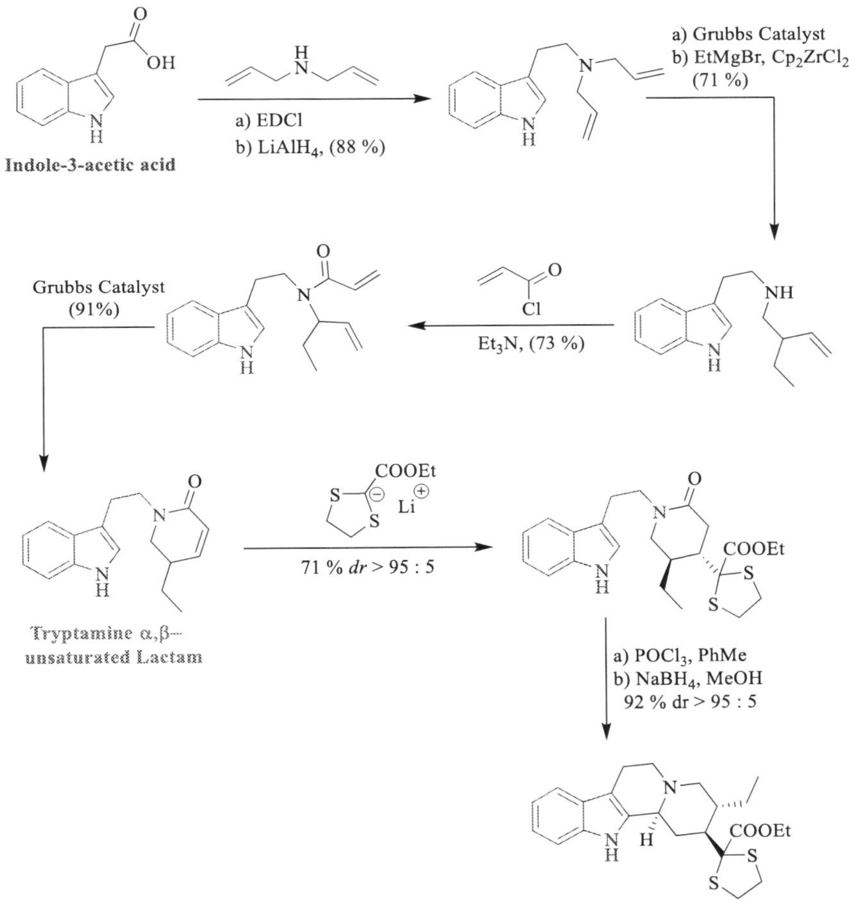

SCHEME 10.7 Martin's racemic formal synthesis of (−)-rhynchophylline and (+)-isorhynchophylline via two ring-closing metathesis (RCM) reactions.[11]

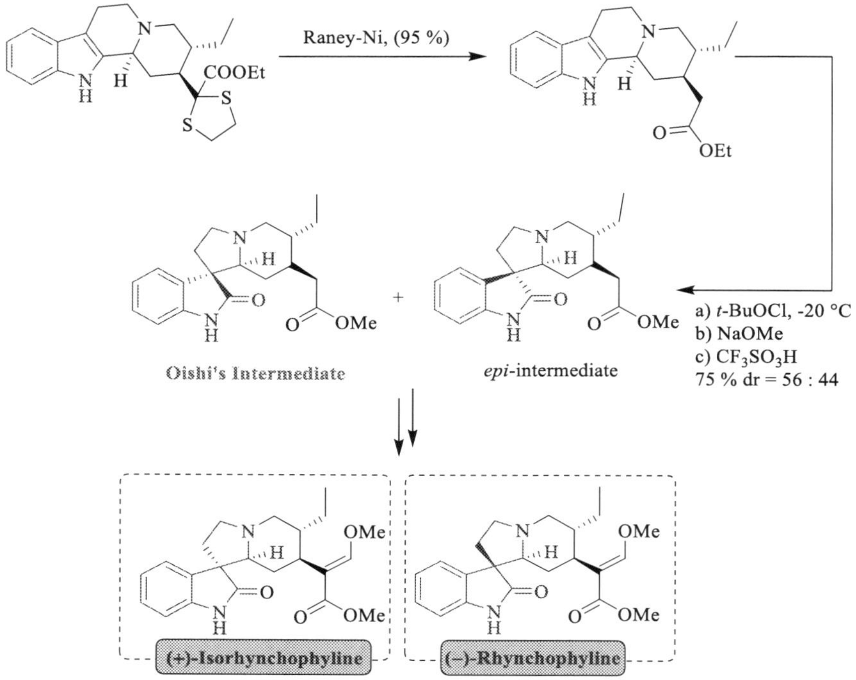

162 — 10. Rhynchophylline and isorhynchophylline

SCHEME 10.7 (Continued).

Key features

- Amide coupling on indole-3-acetic acid
- One-pot ring-closing metathesis (RCM)-carbomagnesation sequence
- Second RCM to furnish α,β-unsaturated lactam
- Diastereoselective 1,4-additions to α,β-unsaturated lactams to furnish side chain

10.3.2 Itoh's enantioselective formal synthesis of (−)-isorhynchophylline via Mannich−Michael reaction (2010)

See Scheme 10.8.

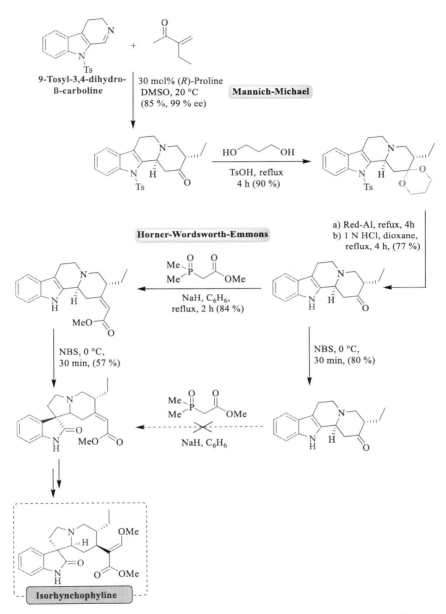

SCHEME 10.8 Itoh's enantioselective formal synthesis of isorhynchophylline via Mannich–Michael reaction.[12]

10. Rhynchophylline and isorhynchophylline

Key features

- Enantioselective Mannich—Michael addition of α,β-unsaturated ketones to *N*-tosyl dihydrocarboline derivative
- Protection of carbonyl group and removal of *N*-tosyl group
- Horner—Wordsworth—Emmons reaction
- Stereoselective oxidative construction of spiro-ring

10.3.3 Wang's enantioselective formal synthesis (2013)

See Scheme 10.9.

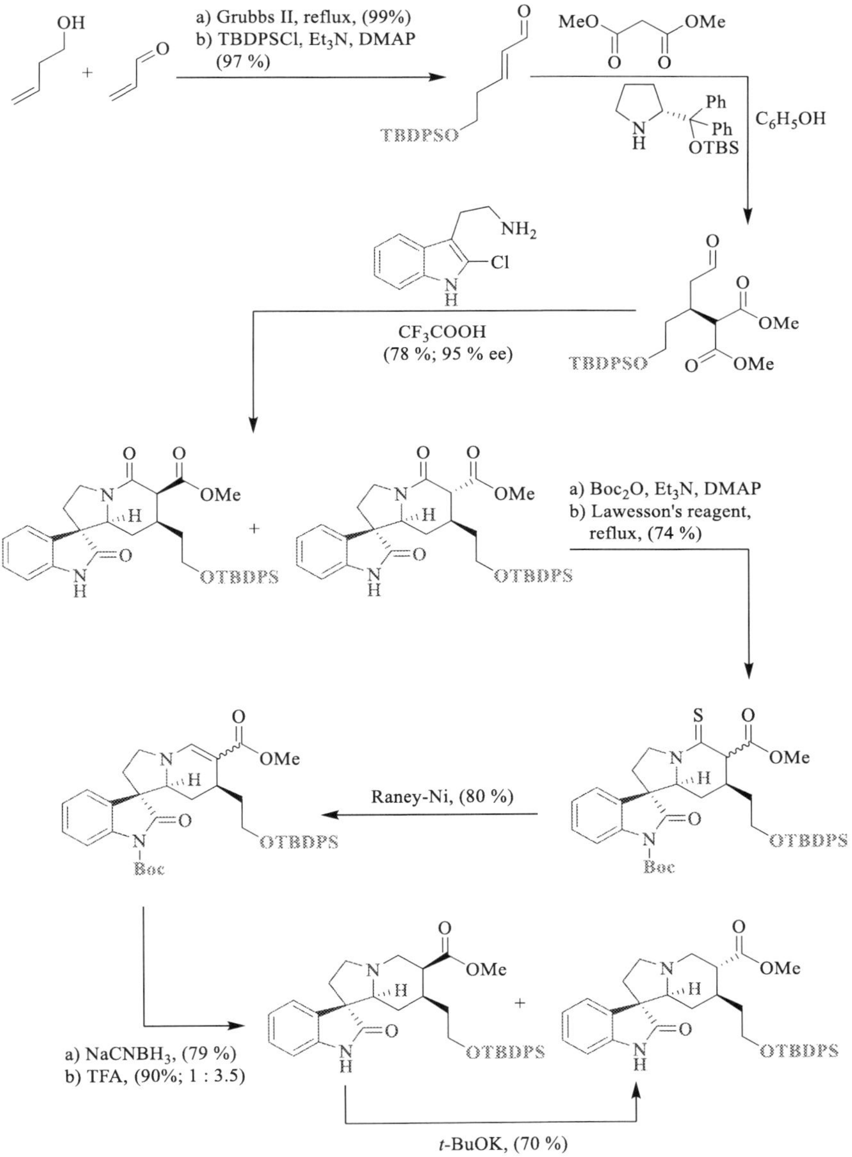

SCHEME 10.9 Wang's enantioselective formal synthesis of (−)-rhynchophylline and (+)-isorhynchophylline.[13]

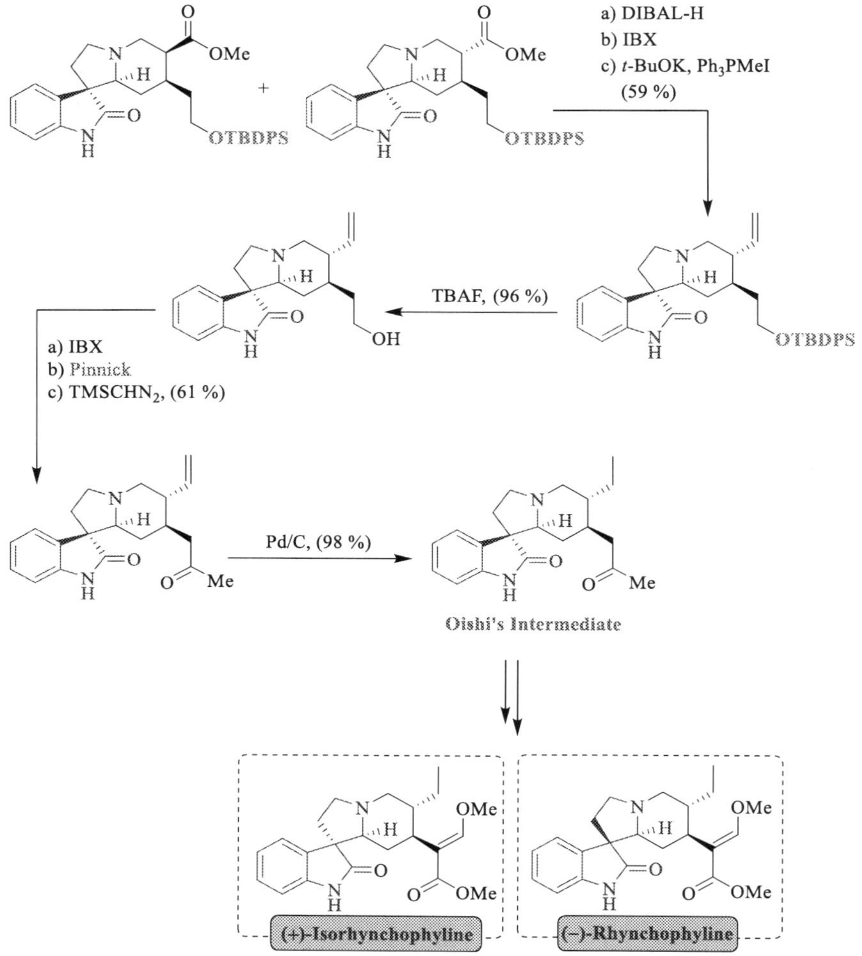

10. Rhynchophylline and isorhynchophylline

a) DIBAL-H
b) IBX
c) *t*-BuOK, Ph₃PMeI
(59 %)

TBAF, (96 %)

a) IBX
b) Pinnick
c) TMSCHN₂, (61 %)

Pd/C, (98 %)

Oishi's Intermediate

(+)-Isorhynchophyline

(−)-Rhynchophyline

SCHEME 10.9 (Continued).

Indole Alkaloids

Key features

- Cross-metathesis of acroleine and 3-butenol
- Hydroxyl group protection of resultant compound
- Asymmetric organocatalytic Michael addition of dialkyl malonate to α,β-unsaturated aldehyde
- Condensation of Michael addition resultant product with 2-halotryptamine to furnish key spiral tetracyclic core
- Conversion of lactam ester to thiolactam and subsequent reductions
- Chemoselective reduction of ester, deprotection, Pinnick oxidation, and esterification of side chain to afford Ban's intermediate

10.3.4 Amat's enantioselective formal synthesis (2013)

See Scheme 10.10.

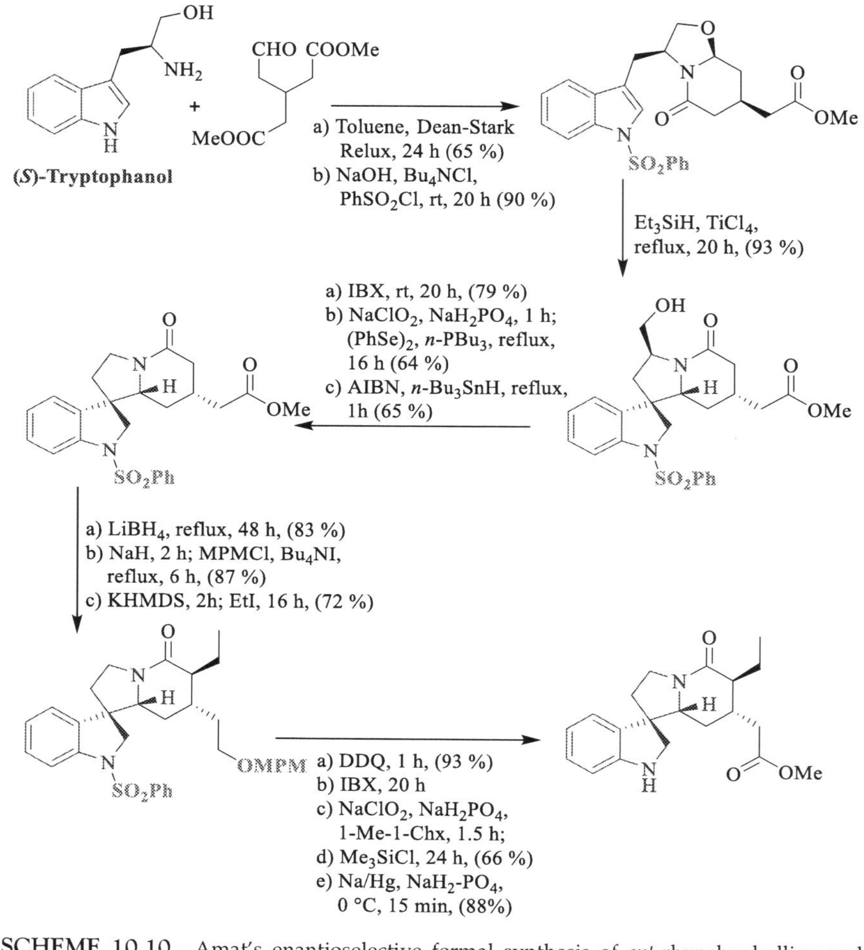

SCHEME 10.10 Amat's enantioselective formal synthesis of *ent*-rhynchophylline and *ent*-isorhynchophylline.[14]

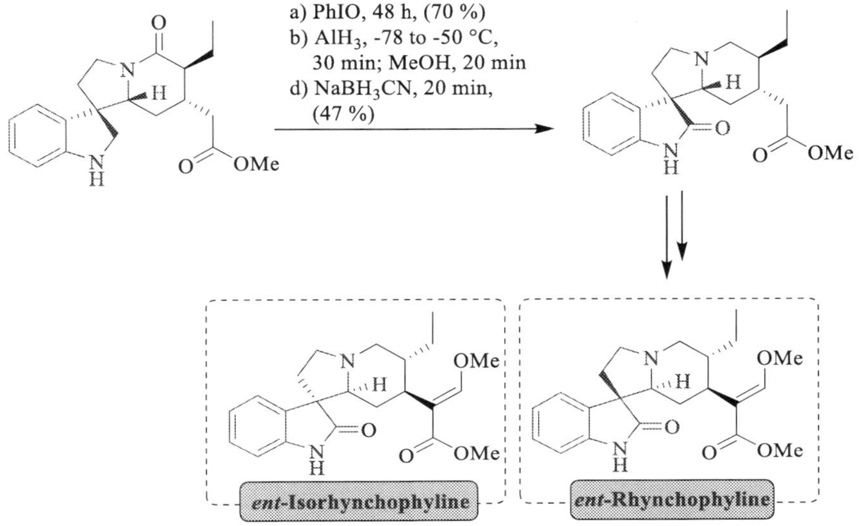

10.3 Formal syntheses of (−)-rhynchophylline and (+)-isorhynchophylline **169**

Reagents:
a) PhIO, 48 h, (70 %)
b) AlH₃, -78 to -50 °C,
   30 min; MeOH, 20 min
d) NaBH₃CN, 20 min,
   (47 %)

ent-Isorhynchophyline

ent-Rhynchophyline

SCHEME 10.10 (Continued).

Key features

- Stereoselective cyclocondensation on (*S*)-tryptophanol
- Stereoselective spirocyclization
- Removal of −CH₂OH group from spiral five membered ring
- Stereoselective alkylation
- Oxidation of indole to oxindole
- Chemoselective carbonyl reduction
- Methoxyvinylation to side chain of spirooxindole intermediate

10.3.5 Xia's racemic formal synthesis via one-pot N-alkylation/CDC and one-pot Michael–Krapcho sequence (2016)[15]

See Scheme 10.11.

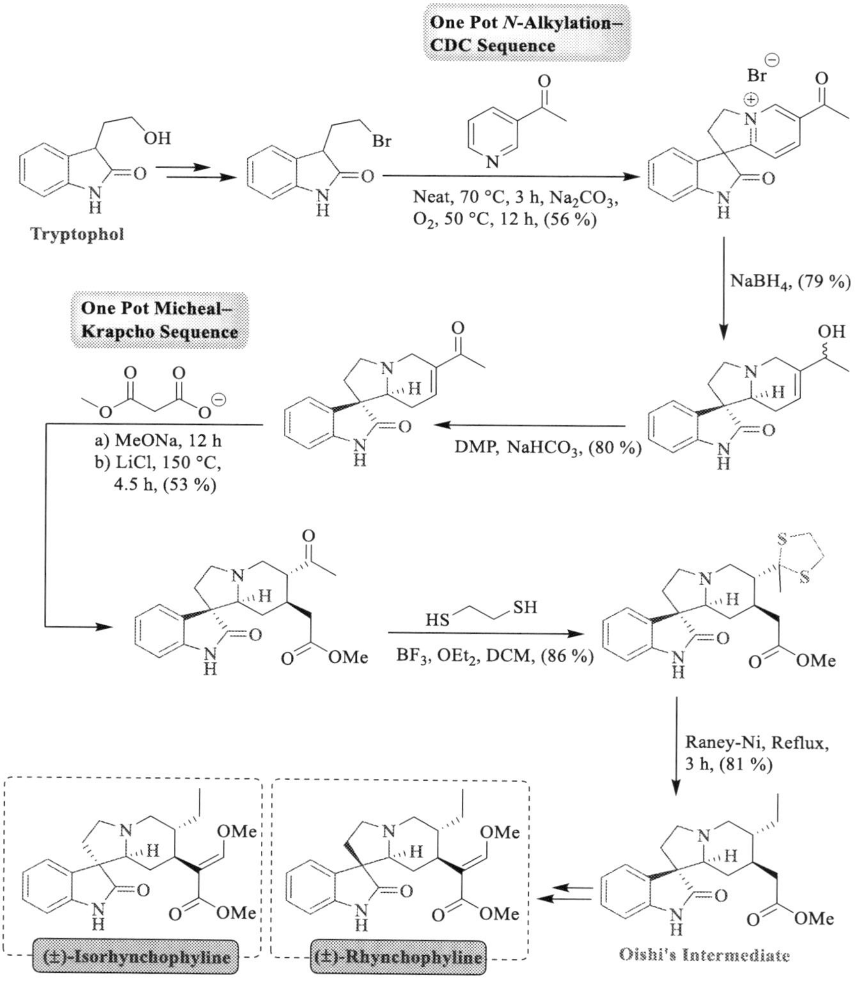

SCHEME 10.11 Xia's racemic formal synthesis of (±)-rhynchophylline and (±)-isorhynchophylline.[15]

Key features

- One-pot N-alkylation/cross-dehydrogenative coupling (CDC) sequence

- Dearomatization
- One-pot Michael—Krapcho sequence
- Diastereoselective hydrogenation of carbonyl group to furnish Ban's intermediate

References

1. Endo, K.; Oshima, Y.; Kikuchi, H.; Koshihara, Y.; Hikino, H. Hypotensive Principles of Uncaria Hooks. *Planta Med.* **1983,** *49* (11), 188−190.
2. Kang, T.-H.; Murakami, Y.; Takayama, H.; Kitajima, M.; Aimi, N.; Watanabe, H.; Matsumoto, K. Protective Effect of Rhynchophylline and Isorhynchophylline on In Vitro Ischemia-Induced Neuronal Damage in the Hippocampus: Putative Neurotransmitter Receptors Involved in Their Action. *Life Sci.* **2004,** *76* (3), 331−343.
3. Ng, Y. P.; Or, T. C. T.; Ip, N. Y. Plant Alkaloids as Drug Leads for Alzheimer's Disease. *Neurochem. Int.* **2015,** *89*, 260−270.
4. Fu, A. K.; Hung, K.-W.; Huang, H.; Gu, S.; Shen, Y.; Cheng, E. Y.; Ip, F. C.; Huang, X.; Fu, W.-Y.; Ip, N. Y. Blockade of EphA4 Signaling Ameliorates Hippocampal Synaptic Dysfunctions in Mouse Models of Alzheimer's Disease. *Proc. Natl. Acad. Sci. U.S.A.* **2014,** *111* (27), 9959−9964.
5. Ban, Y.; Seto, M.; Oishi, T. Stereoselective Total Synthesis of (±)-Rhynchophylline and (±)-Isorhynchophylline. *Tetrahedron Lett.* **1972,** *13* (21), 2113−2116.
6. Ban, Y.; Seto, M.; Oishi, T. The Synthesis of 3-Spirooxindole Derivatives. VII. Total Synthesis of Alkaloids (±)-Rhynchophylline and (±)-Isorhynchophylline. *Chem. Pharm. Bull.* **1975,** *23* (11), 2605−2613.
7. Wanner, M. J.; Ingemann, S.; van Maarseveen, J. H.; Hiemstra, H. Total Synthesis of the Spirocyclic Oxindole Alkaloids Corynoxine, Corynoxine B, Corynoxeine, and Rhynchophylline. *Eur. J. Org. Chem.* **2013,** *2013* (6), 1100−1106.
8. Zhang, Z.; Zhang, W.; Kang, F.; Ip, F. C.; Ip, N. Y.; Tong, R. Asymmetric Total Syntheses of Rhynchophylline and Isorhynchophylline. *J. Org. Chem.* **2019,** *84* (17), 11359−11365.
9. Finch, N.; Taylor, W. Oxidative Transformations of Indole Alkaloids. I. The Preparation of Oxindoles From Yohimbine; The Structures and Partial Syntheses of Mitraphylline, Rhyncophylline and Corynoxeine. *J. Am. Chem. Soc.* **1962,** *84* (20), 3871−3877.
10. Brown, R.; CL, C. A Novel Synthesis of Isorhynchophylline and Rhynchophylline From Secologanin. **1976**.
11. Deiters, A.; Pettersson, M.; Martin, S. F. General Strategy for the Syntheses of Corynanthe, Tacaman, and Oxindole Alkaloids. *J. Org. Chem.* **2006,** *71* (17), 6547−6561.
12. Nagata, K.; Ishikawa, H.; Tanaka, A. Formal Syntheses of Dihydrocorynantheine and Isorhynchophylline via Proline Catalyzed Mannich-Michael Reaction. *Heterocycles* **2010,** *81* (8), 1791−1798.
13. Zhang, H.; Ma, X.; Kang, H.; Hong, L.; Wang, R. The Enantioselective Formal Synthesis of Rhynchophylline and Isorhynchophylline. *Chem. Asian J.* **2013,** *8* (3), 542−545.
14. Amat, M.; Ramos, C.; Pérez, M.; Molins, E.; Florindo, P.; Santos, M. M.; Bosch, J. Enantioselective Formal Synthesis of Ent-Rhynchophylline and Ent-Isorhynchophylline. *Chem. Commun.* **2013,** *49* (19), 1954−1956.
15. Xu, J.; Shao, L.-D.; Shi, X.; Ren, J.; Xia, C.; Zhao, Q.-S. Collective Formal Synthesis of (±)-Rhynchophylline and Homologues. *RSC Adv.* **2016,** *6* (68), 63131−63135.

CHAPTER 11

Spirotryprostatin A

Spirotryprostatin A (Fig. 11.1) was first isolated in 1996 by Osada and coworkers[1] from the fermentation broth of *Aspergillus fumigatus*. It comprises an annulated diketopiperazine core substituted multiply with proline moiety connected to a 6′-methoxyindolinone via the spiro quaternary stereogenic center.

This molecule is a promising antitumor drug candidate as observed by their inhibition potential of mammalian cell cycle progression at micromolar concentrations in the G2/M phase.[1]

A variety of synthetic methods of spirotryprostatin A have been developed so far. Moreover, different studies have also been done to explore the structurally similar molecules to discover the bioactivity of the natural product-inspired framework. Various protocols developed so far include intramolecular cyclization of iminium ion, oxidative rearrangement of chiral tetrahydro-β-carbolines, intramolecular Heck reactions, [3 + 2] and [5 + 2] cycloadditions, Mannich reactions, nitroolefination, Pd-catalyzed prenylation, and magnesium iodide-catalyzed ring-expansion that are detailed described in this section.

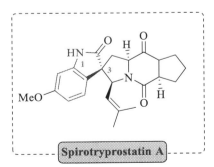

FIGURE 11.1   Structure of spirotryprostatin A.

174

11. Spirotryprostatin A

11.1 Total synthesis of spirotryprostatin A

11.1.1 Danishefsky's total synthesis of spirotryprostatin A (1998, 1999)

See Scheme 11.1.

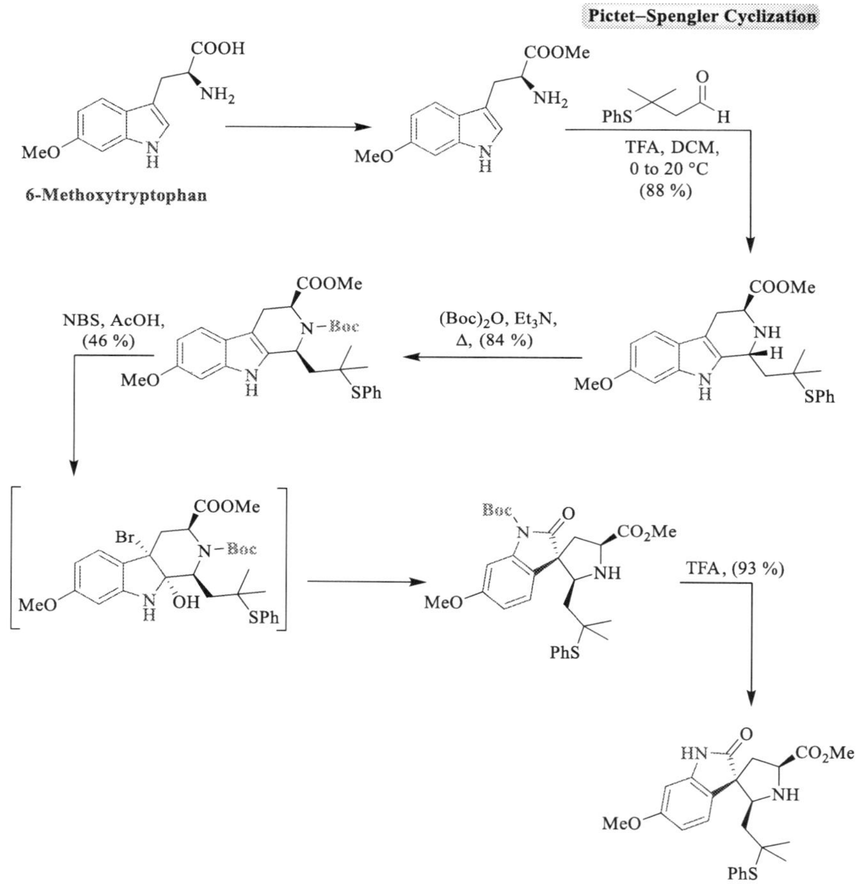

SCHEME 11.1 Danishefsky's total synthesis of spirotryprostatin A.[2,3]

Indole Alkaloids

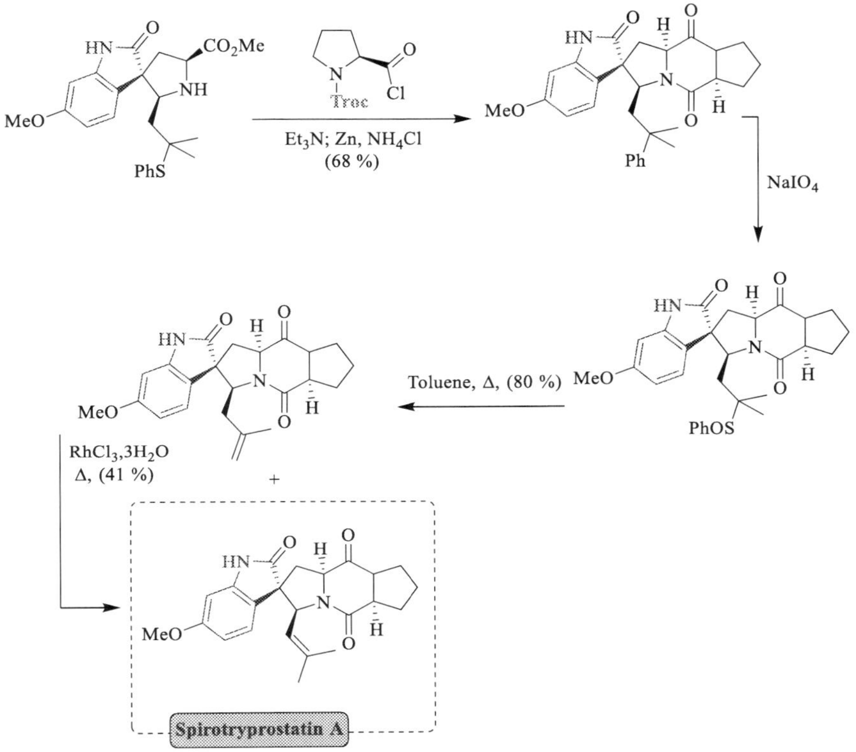

11.1 Total synthesis of spirotryprostatin A 175

SCHEME 11.1 (Continued).

Key features

- Pictet–Spengler reaction of aldehyde and 6-methoxytryptophan derivative
- Oxidative rearrangement of the β-carboline derivative to oxindole by the action of *N*-bromosuccinimide
- Introduction of diketopiperazine linkage to synthesized spirooxindole

11.1.2 Williams' asymmetric total synthesis of spirotryprostatin A (2003, 2004)

See Scheme 11.2.

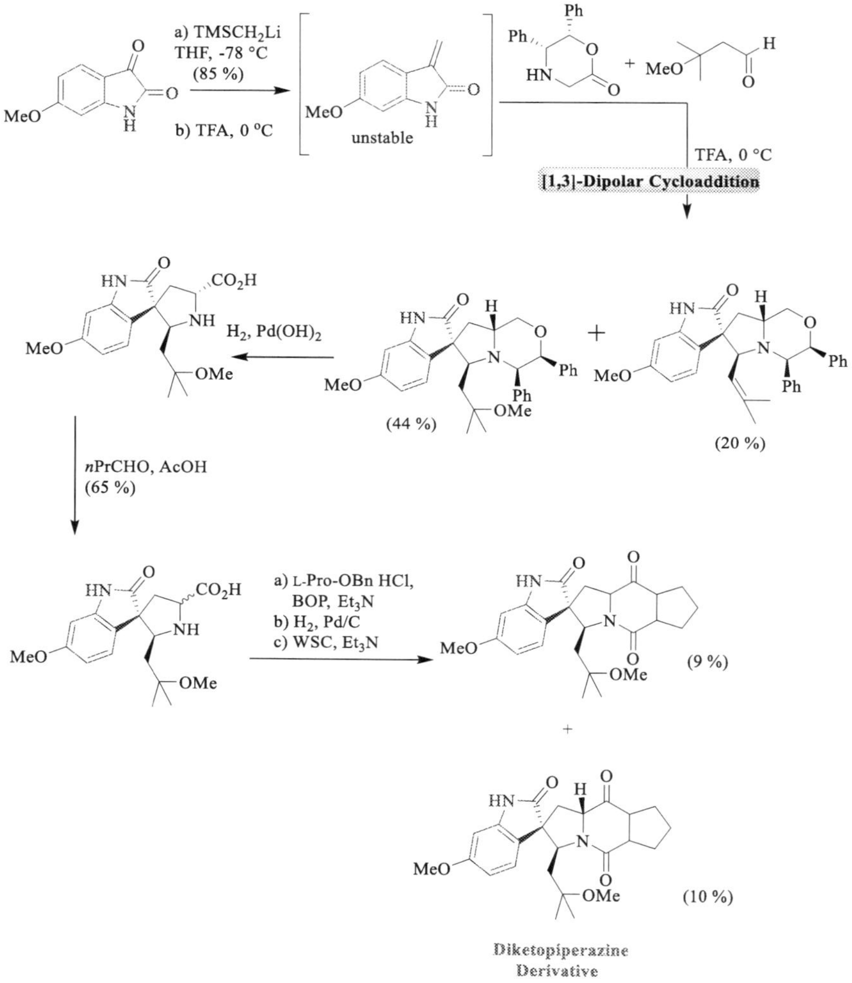

SCHEME 11.2 Williams' asymmetric total synthesis of spirotryprostatin A.[4,5]

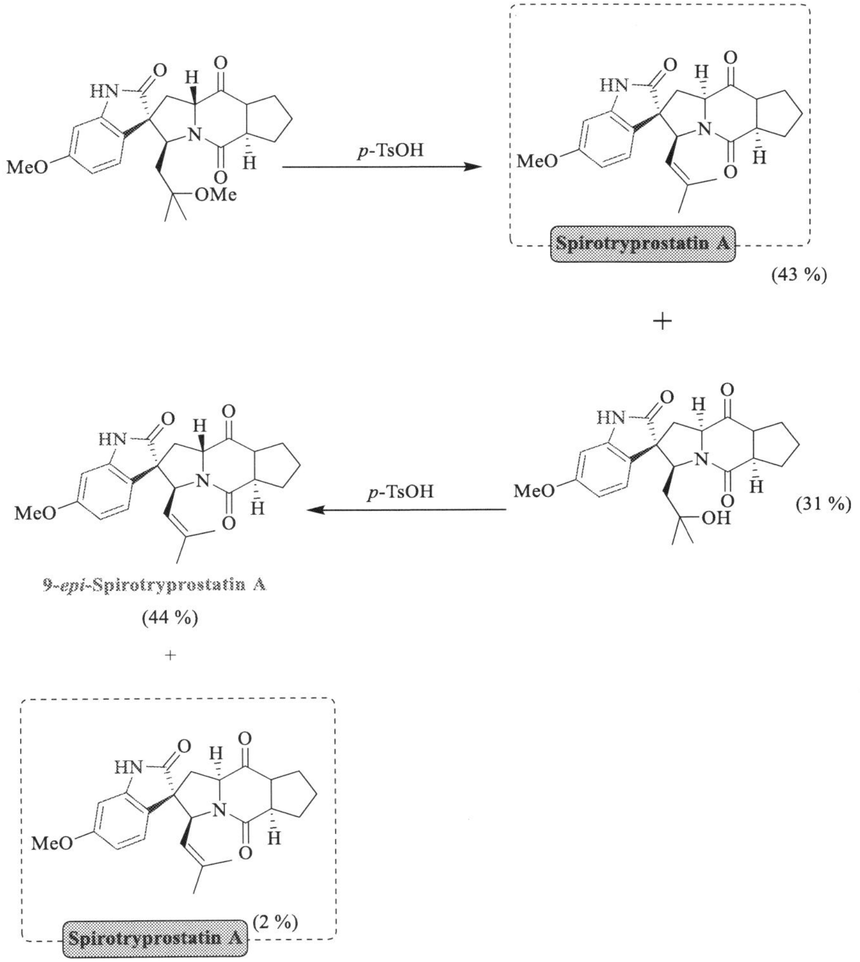

11.1 Total synthesis of spirotryprostatin A   **177**

SCHEME 11.2   (Continued).

Key features

- Asymmetric total synthesis of spirotryprostatin A in 12 steps (7 steps in the longest linear sequence) from commercially available reagents.
- Key step is asymmetric [1,3]-dipolar cycloaddition reaction of methylene indolinone.

178

11. Spirotryprostatin A

11.1.3 Fukuyama's stereoselective total synthesis of spirotryprostatin A (2014)

See Scheme 11.3.

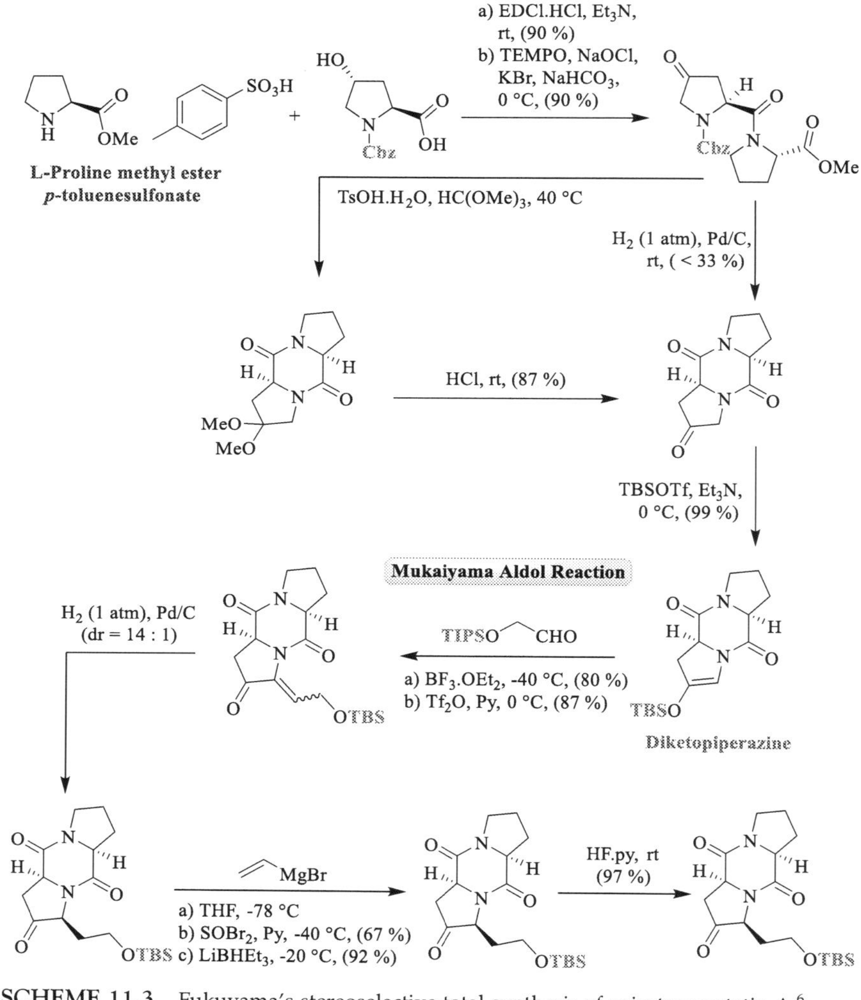

SCHEME 11.3 Fukuyama's stereoselective total synthesis of spirotryprostatin A.[6]

Indole Alkaloids

11.1 Total synthesis of spirotryprostatin A

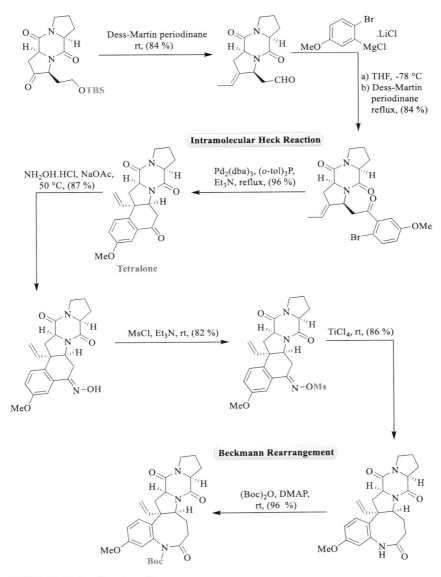

SCHEME 11.3 (Continued).

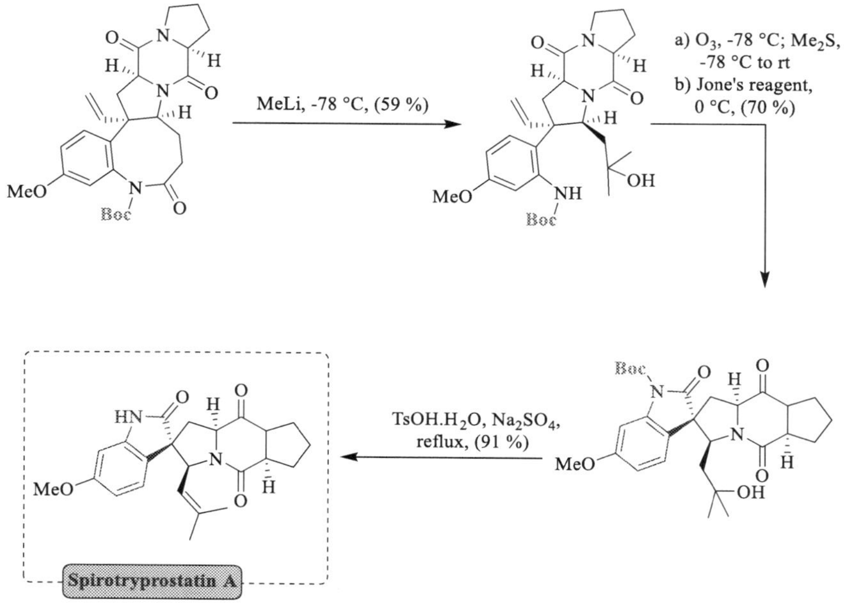

11. Spirotryprostatin A

SCHEME 11.3 (Continued).

Key features

- Condensation of two L-proline derivatives to furnish *cyclo*-(Pro–Pro) diketopiperazine
- Introduction of an alkyl group at C18 position with desired (*S*)-stereochemistry via Mukaiyama aldol reaction of *cyclo*-(Pro–Pro) diketopiperazine with silyloxy acetaldehyde
- Stereo-controlled Intramolecular Heck reaction with a carbonyl tether
- Beckmann rearrangement to construct spiro-connection

11.1.4 Gong's synthesis of diastereoisomers of spirotryprostatin A via asymmetric organocatalytic 1,3-dipolar cycloaddition (2011)

See Scheme 11.4.

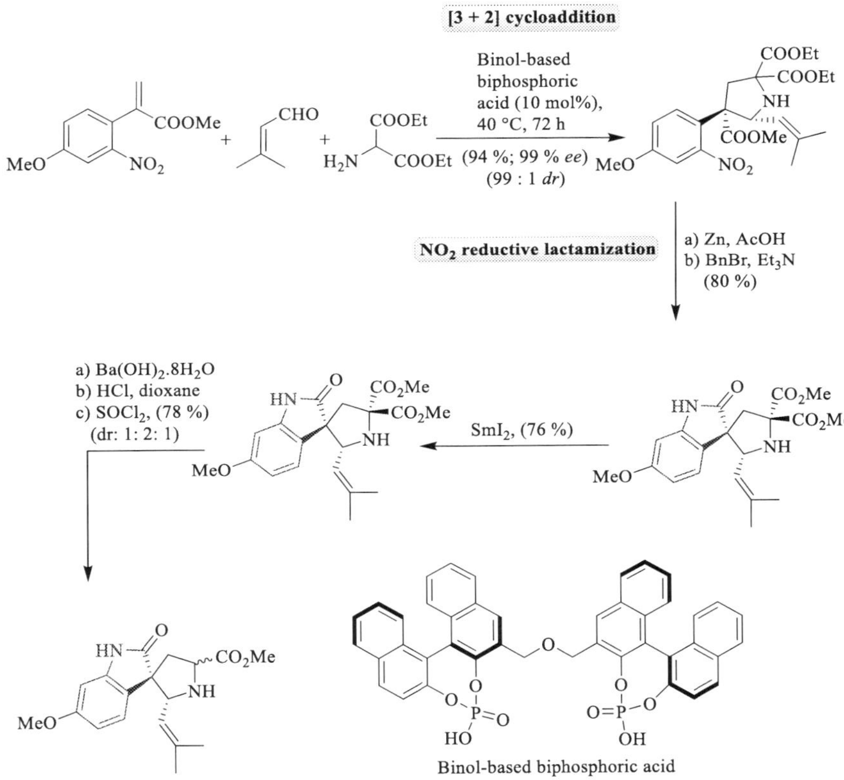

SCHEME 11.4 Gong's synthesis of diastereoisomers of spirotryprostatin A via asymmetric organocatalytic 1,3-dipolar cycloaddition.[7]

Key features

- Chiral Bronsted acid-catalyzed [3 + 2] cycloaddition of azomethine ylides
- Nitro reductive lactamization to yield spirooxindole
- Structural changes via reduction, protection, deprotection, hydrolysis, decarboxylation, esterification, and then lactamization to furnish diastereoisomers of spirotryprostatin A

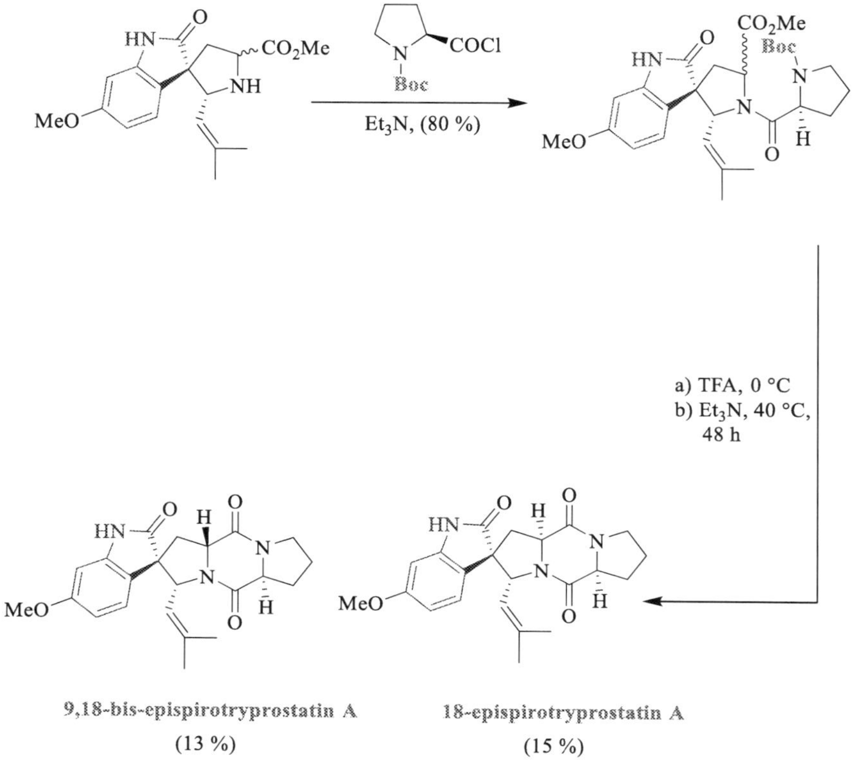

182

11. Spirotryprostatin A

SCHEME 11.4 (Continued).

References

1. Cui, C.-B.; Kakeya, H.; Osada, H. Novel Mammalian Cell Cycle Inhibitors, Spirotryprostatins A and B, Produced by *Aspergillus fumigatus*, Which Inhibit Mammalian Cell Cycle at G2/M Phase. *Tetrahedron* **1996**, *52* (39), 12651–12666.
2. Edmondson, S. D.; Danishefsky, S. J. The Total Synthesis of Spirotryprostatin A. *Angew. Chem. Int. Ed.* **1998**, *37* (8), 1138–1140.
3. Edmondson, S.; Danishefsky, S. J.; Sepp-Lorenzino, L.; Rosen, N. Total Synthesis of Spirotryprostatin A, Leading to the Discovery of Some Biologically Promising Analogues. *J. Am. Chem. Soc.* **1999**, *121* (10), 2147–2155.
4. Onishi, T.; Sebahar, P. R.; Williams, R. M. Concise, Asymmetric Total Synthesis of Spirotryprostatin A. *Org. Lett.* **2003**, *5* (17), 3135–3137.
5. Onishi, T.; Sebahar, P. R.; Williams, R. M. Concise, Asymmetric Total Synthesis of Spirotryprostatin A. *Tetrahedron* **2004**, *60* (42), 9503–9515.
6. Kitahara, K.; Shimokawa, J.; Fukuyama, T. Stereoselective Synthesis Of Spirotryprostatin A. *Chem. Sci.* **2014**, *5* (3), 904–907.
7. Cheng, M.-N.; Wang, H.; Gong, L.-Z. Asymmetric Organocatalytic 1,3-Dipolar Cycloaddition of Azomethine Ylide to Methyl 2-(2-nitrophenyl) Acrylate for the Synthesis of Diastereoisomers of Spirotryprostatin A. *Org. Lett.* **2011**, *13* (9), 2418–2421.

CHAPTER 12

Spirotryprostatin B

Spirotryprostatin B (Fig. 12.1) was first isolated in 1996 by Osada and coworkers[1] from the fermentation broth of the same fungus *Aspergillus fumigatus* from which spirotryprostatin A was isolated. Structurally, it resembles closely spirotryprostatin A differing only in methoxy substitution at 6-indolinone position of oxindole.

It is also of wide biological importance due to its inhibition potential of mammalian cell cycle progression in the G2/M phase and possesses more efficiency than its saturated congener spirotryprostatin A.[1]

Key step in the total synthesis of spirotryprostatin B is asymmetric stereocontrolled construction of the quaternary C3 and adjacent prenyl-substituted C18 centers of the dehydrospiropyrrolidine ring. A total of eight different approaches for the total synthesis of this molecule have been reported.

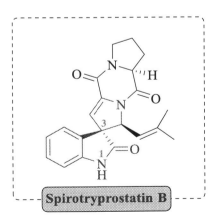

FIGURE 12.1 Structure of (−)-spirotryprostatin B.

184

12. Spirotryprostatin B

12.1 Total syntheses of (−)-spirotryprostatin B

12.1.1 Danishefsky's total synthesis of spirotryprostatin B (2000)

See Scheme 12.1.

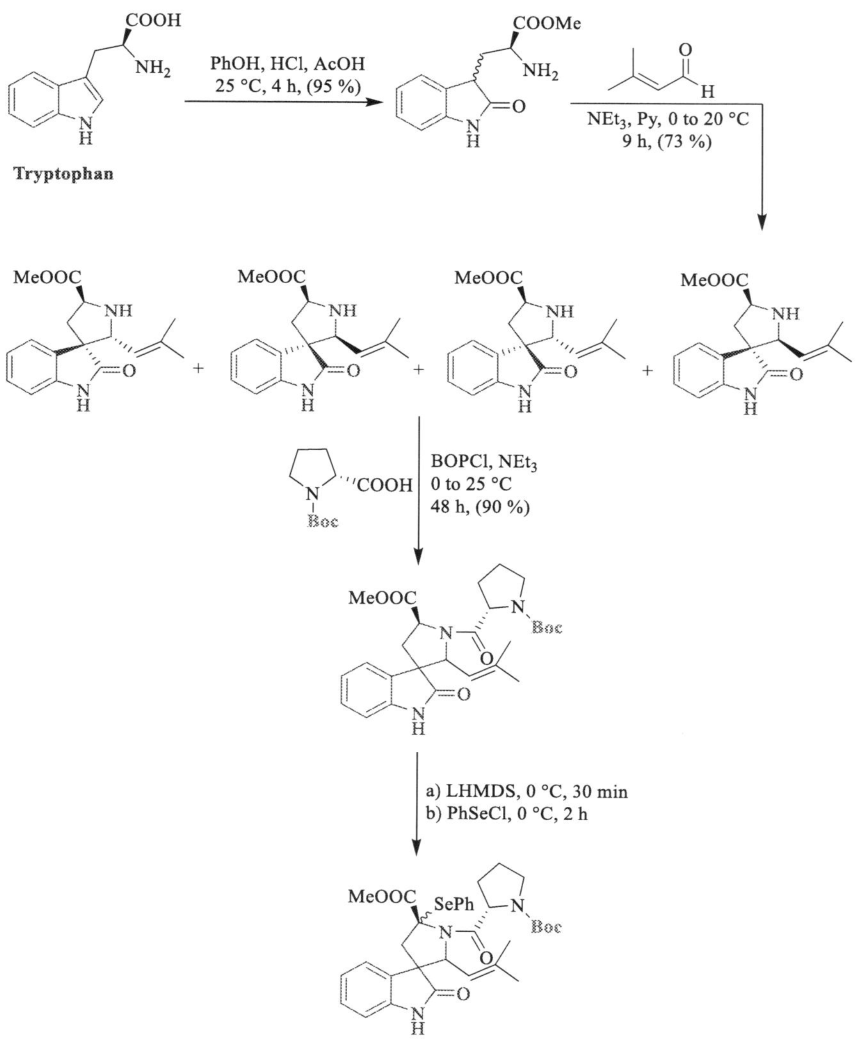

SCHEME 12.1 Danishefsky's total synthesis of spirotryprostatin B.[2]

Indole Alkaloids

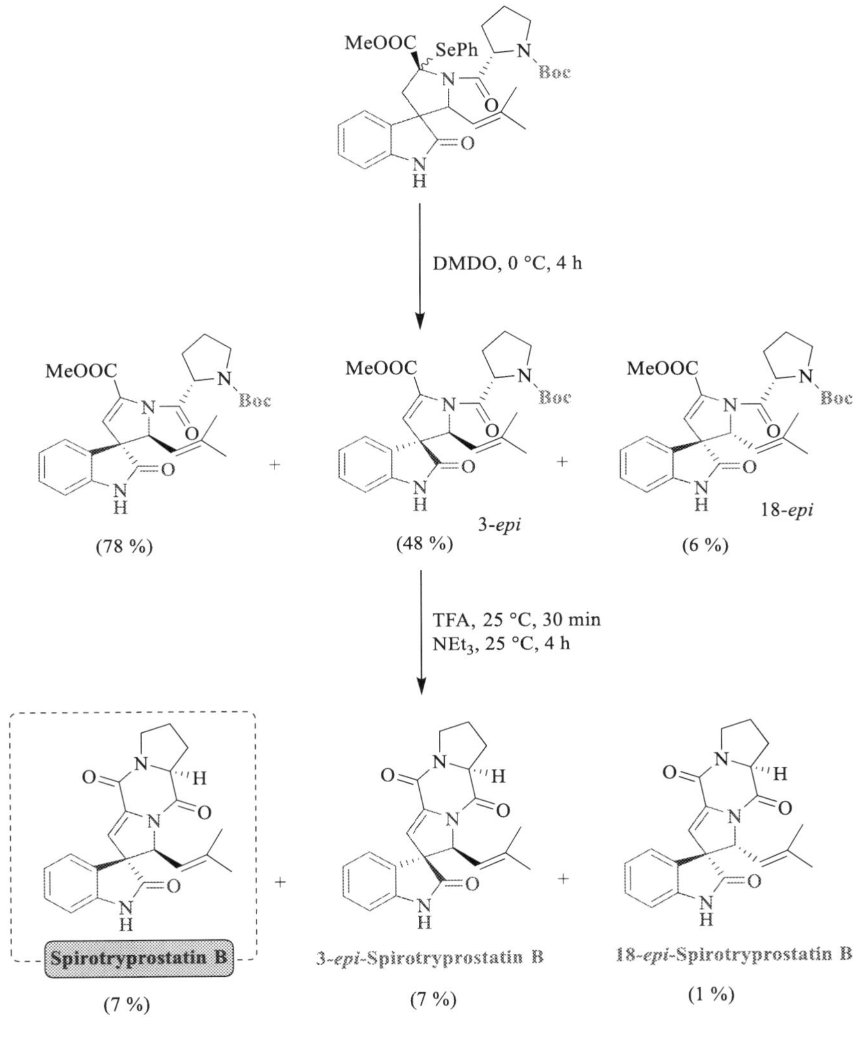
MeOOC
SePh
N
Boc
O
N
O
N
H
DMDO, 0 °C, 4 h
3-epi
18-epi
(78 %)
(48 %)
(6 %)
TFA, 25 °C, 30 min
NEt3, 25 °C, 4 h
H
Spirotryprostatin B
3-epi-Spirotryprostatin B
18-epi-Spirotryprostatin B
(7 %)
(7 %)
(1 %)

SCHEME 12.1 (Continued).

Key features

- Prenyl insertion to oxindole derivative of tryptophan
- Condensation with *N*-Boc-*N*-Boc-L-proline
- Intramolecular cyclization leading to diketopiperazine moiety

186

12. Spirotryprostatin B

12.1.2 Rosen's total synthesis of spirotryprostatin B (2000, 2010)

See Scheme 12.2.

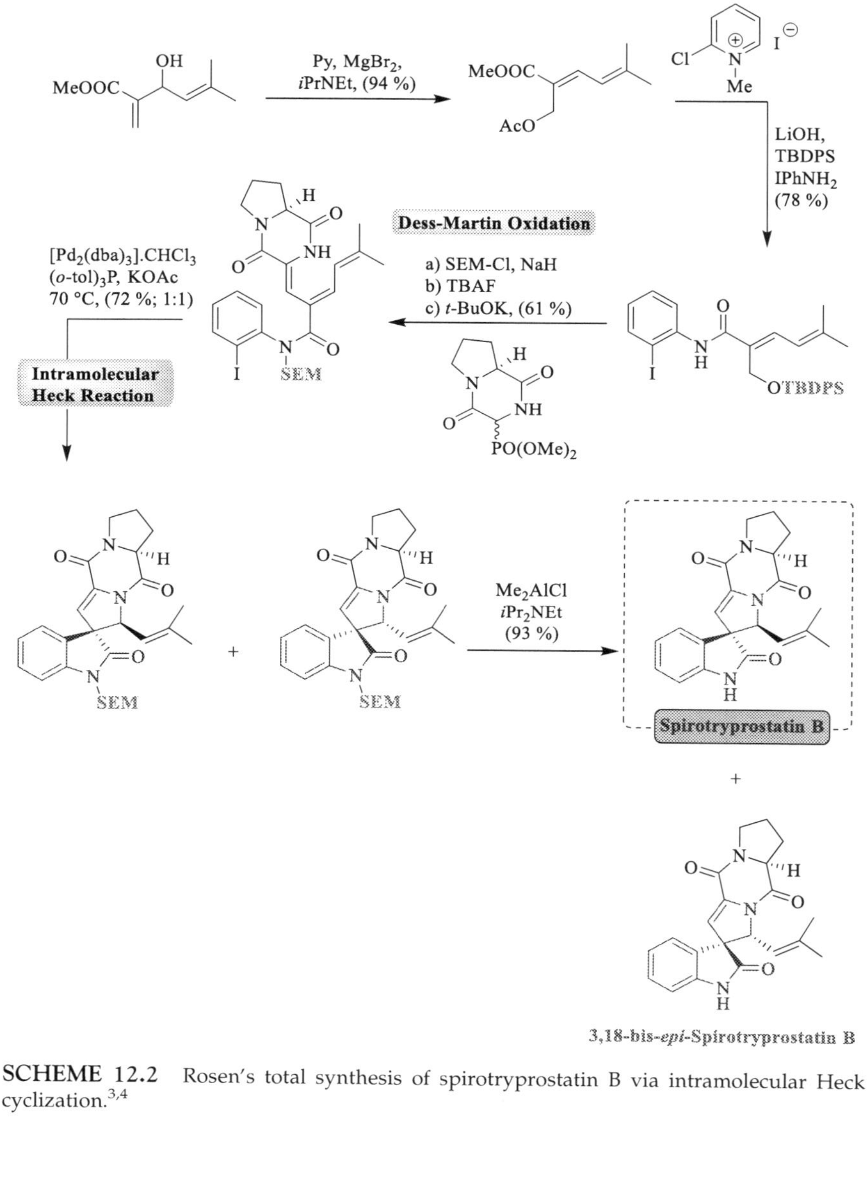

SCHEME 12.2 Rosen's total synthesis of spirotryprostatin B via intramolecular Heck cyclization.[3,4]

Indole Alkaloids

Key features

- Spirotryprostatin B was synthesized in a 9% yield from commercially available precursors by isolating 10 intermediates.
- Key step is catalytic intramolecular Heck reaction, followed by capture of the resulting acyclic, chiral η^3-allylpalladium intermediate by a tethered diketopiperazine.
- (*E*)-Dienoate was synthesized from primary allylic bromide which was formerly prepared from allylic alcohol.
- Formation of (*E*)-dienoate's siloxycarboxylic acid and its coupling with iodoaniline.
- Coupling with diketopiperazine phosphonate.
- Spirocyclization.

12.1.3 Biomimetic total synthesis of spirotryprostatin B by Ganesan (2000)

See Scheme 12.3.

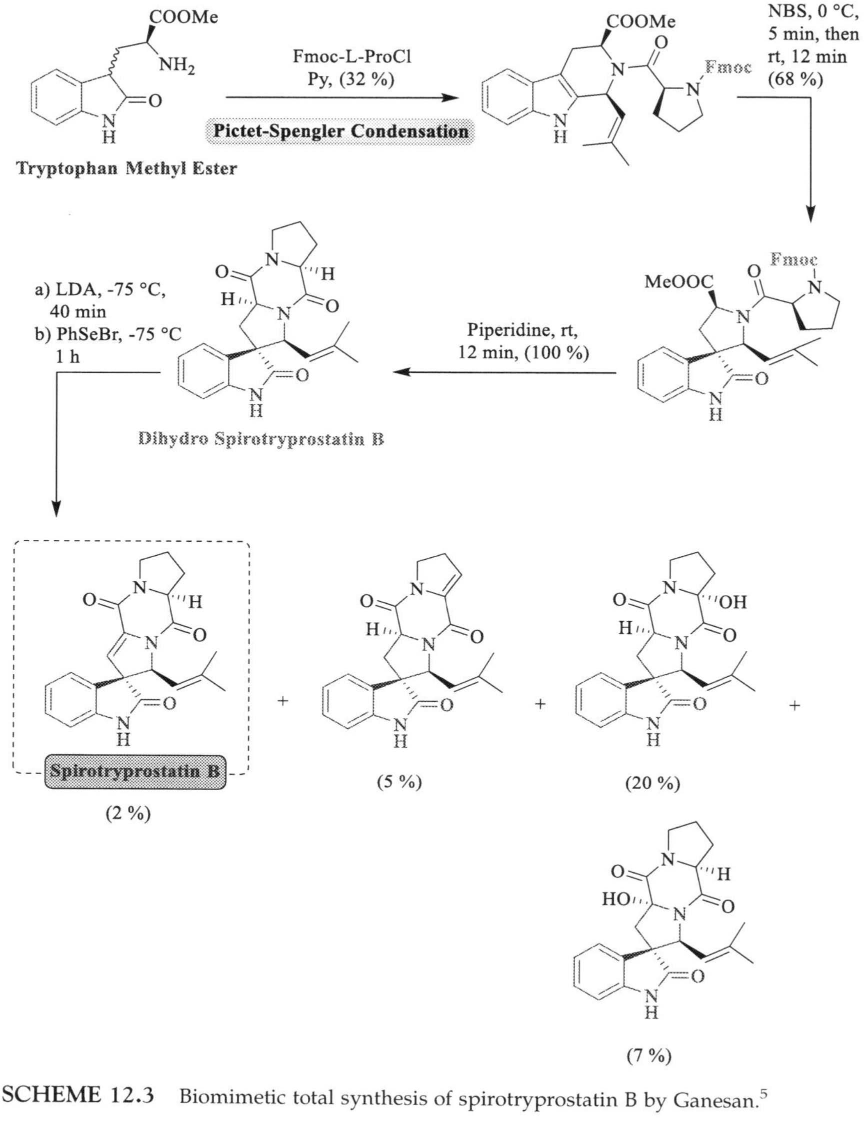

12. Spirotryprostatin B

SCHEME 12.3 Biomimetic total synthesis of spirotryprostatin B by Ganesan.[5]

Indole Alkaloids

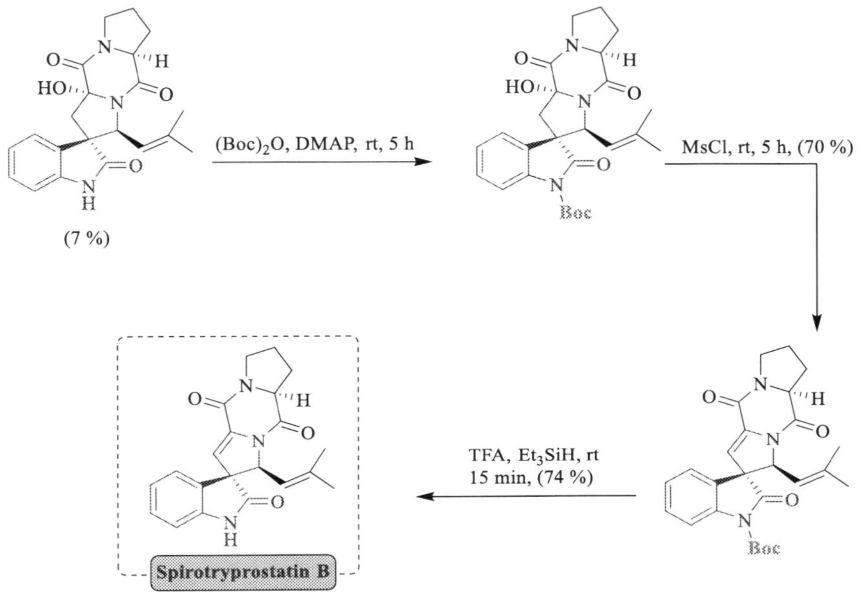

12.1 Total syntheses of (−)-spirotryprostatin B **189**

SCHEME 12.3 (Continued).

Key features

- Pictet−Spengler condensation of tryptophan methyl ester with Fmoc-L-proline
- Spiro ring formation by NBS action
- Synthesis of dihydrospirotryprostatin B
- Unsaturation to furnish spirotryprostatin B

190

12. Spirotryprostatin B

12.1.4 Fuji's total synthesis of spirotryprostatin B via asymmetric nitroolefination (2002)

See Scheme 12.4.

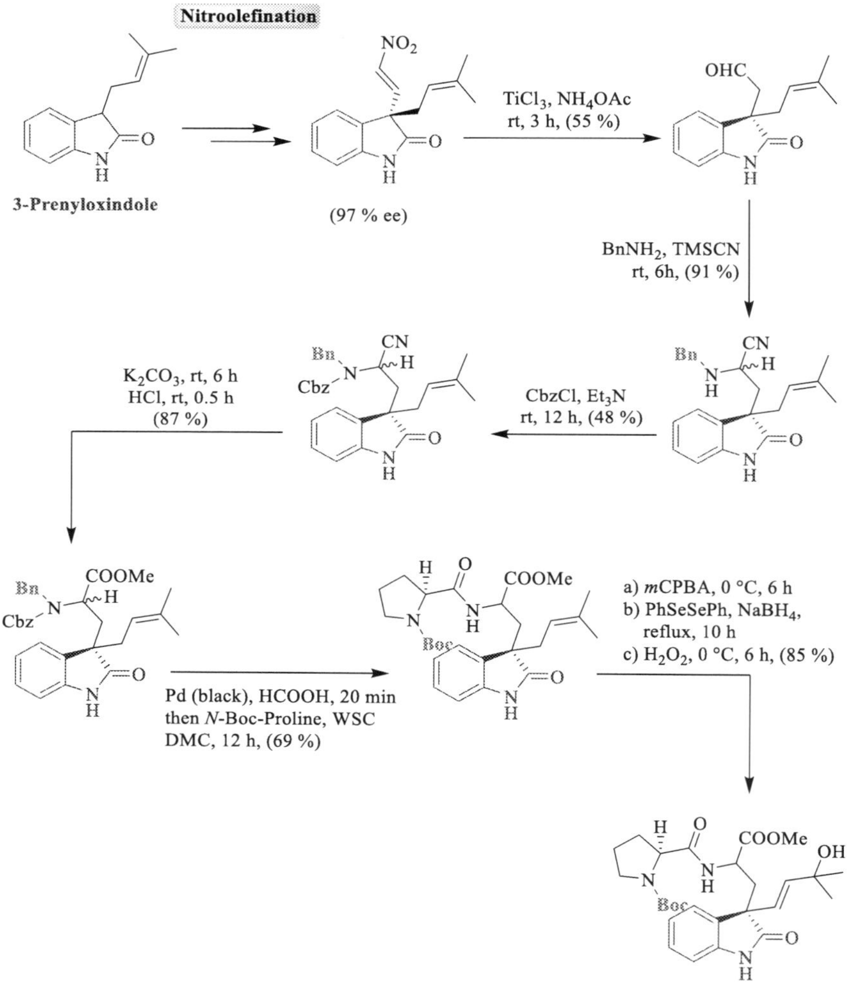

SCHEME 12.4 Fuji's total synthesis of spirotryprostatin B via asymmetric nitroolefination.[6]

Indole Alkaloids

12.1 Total syntheses of (−)-spirotryprostatin B

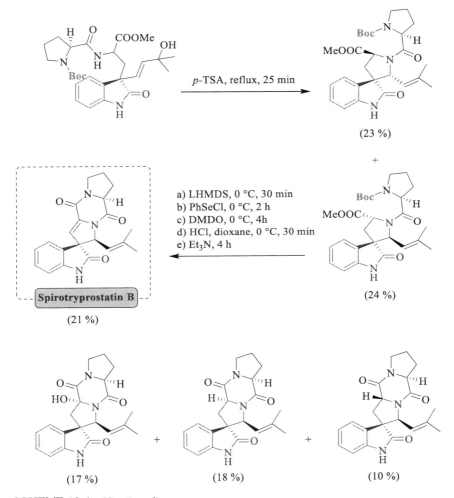

SCHEME 12.4 (Continued).

Key features

- Nitroolefination of 3-prenyloxindole
- Enantioselective installation of a C-3 quaternary stereocenter at amino acid alternate
- Coupling with L-proline
- Spiropyrrolidine ring closure
- Incorporation of a conjugated double bond in the spiropyrrolidine unit
- Deprotection of nitrogen and cyclization to get target molecule

192
12. Spirotryprostatin B

12.1.5 Williams' asymmetric stereocontrolled total synthesis of spirotryprostatin B (2000, 2002)

See Scheme 12.5.

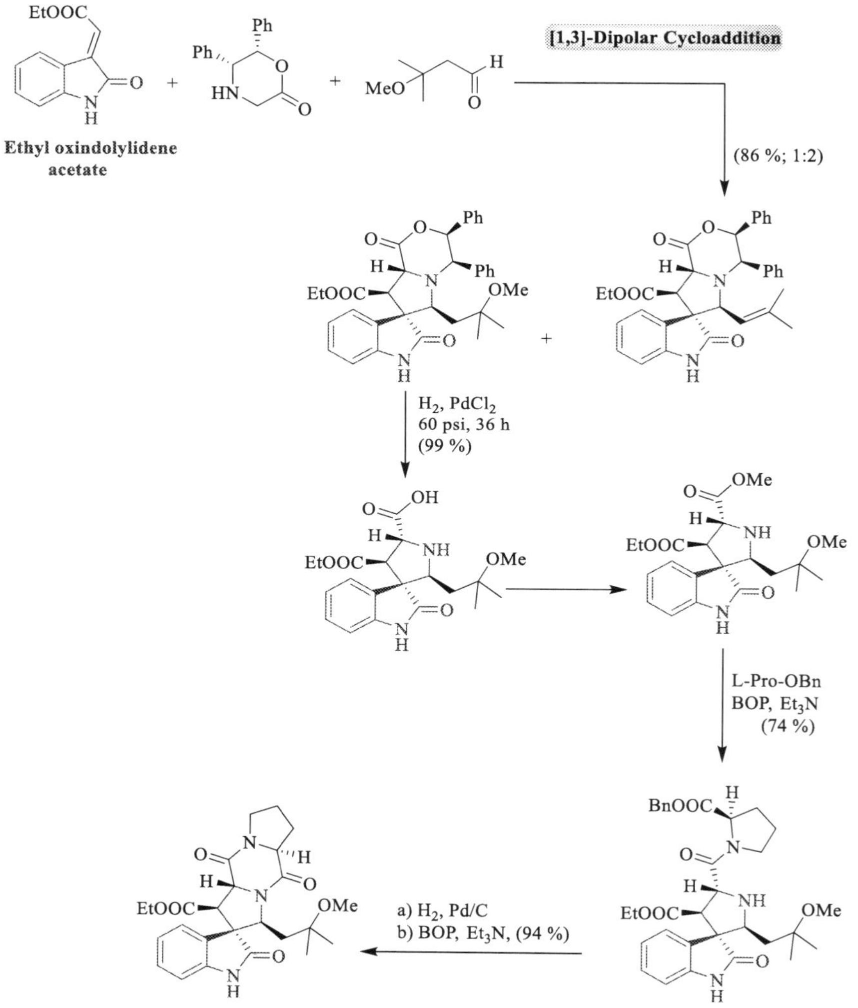

SCHEME 12.5 Williams' asymmetric stereocontrolled total synthesis of spirotryprostatin B.[7,8]

Indole Alkaloids

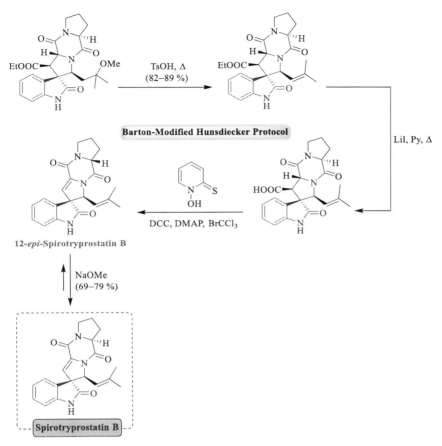

SCHEME 12.5 (Continued).

Key features

- Diastereoselective asymmetric [1,3]-dipolar cycloaddition of azomethine ylide to furnish core structure
- Reductive cleavage of the oxazinone to generate spirooxindole pyrrolidine that was coupled to D-proline benzyl ester and cyclized to the pentacyclic diketopiperazine
- Oxidative decarboxylation via Barton-modified Hunsdiecker protocol to yield 12-*epi*-spirotryprostatin B
- Thermodynamic epimerization of D-proline stereogenic center to yield spirotryprostatin B

12.1.6 Horne's total synthesis of spirotryprostatin B (2004)

See Scheme 12.6.

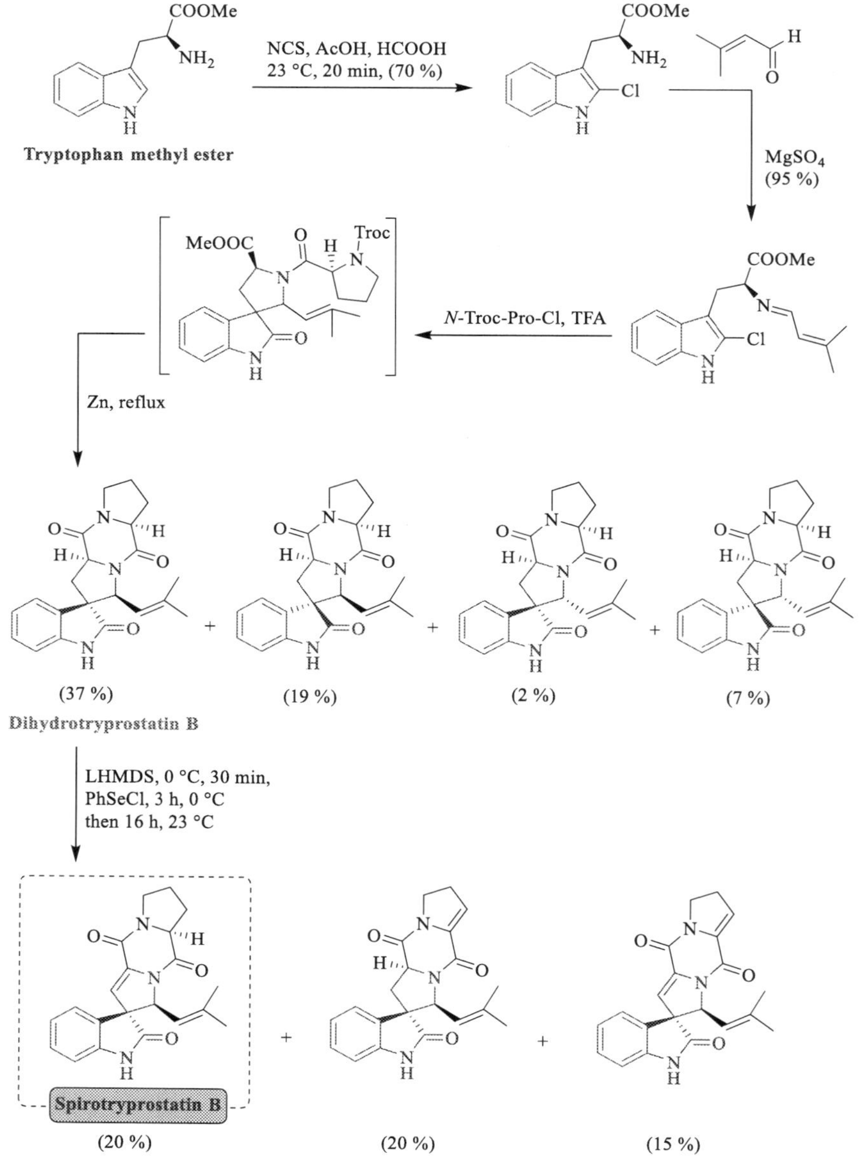

SCHEME 12.6 Horne's total synthesis of spirotryprostatin B.[9]

12.1 Total syntheses of (−)-spirotryprostatin B

Key features

- Synthesis of 2-chlorotryptophan
- Spirocyclization of 2-chlorotryptophan with prenyl aldehyde and insertion of diketopiperazine moiety
- Nonoxidative dehydrogenation of dihydrospirotryprostatin B

12.1.7 Carreira's total synthesis of spirotryprostatin B (2003, 2005)

See Scheme 12.7.

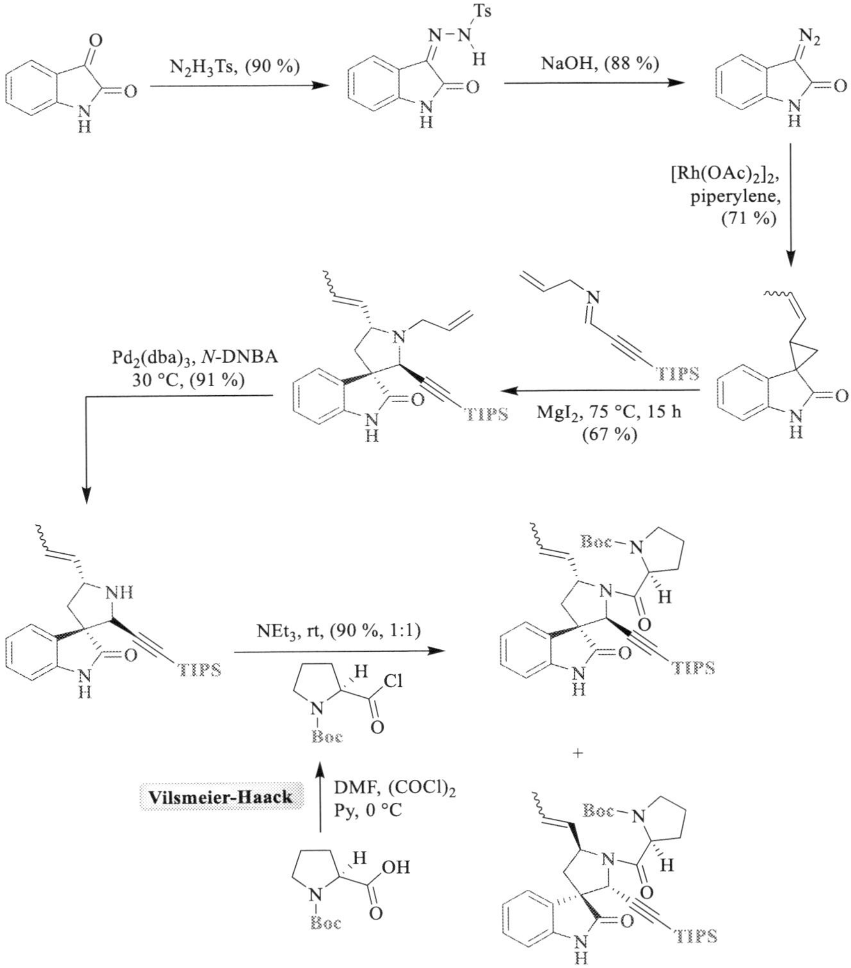

SCHEME 12.7 Carreira's total synthesis of spirotryprostatin B3.[10,11]

Indole Alkaloids

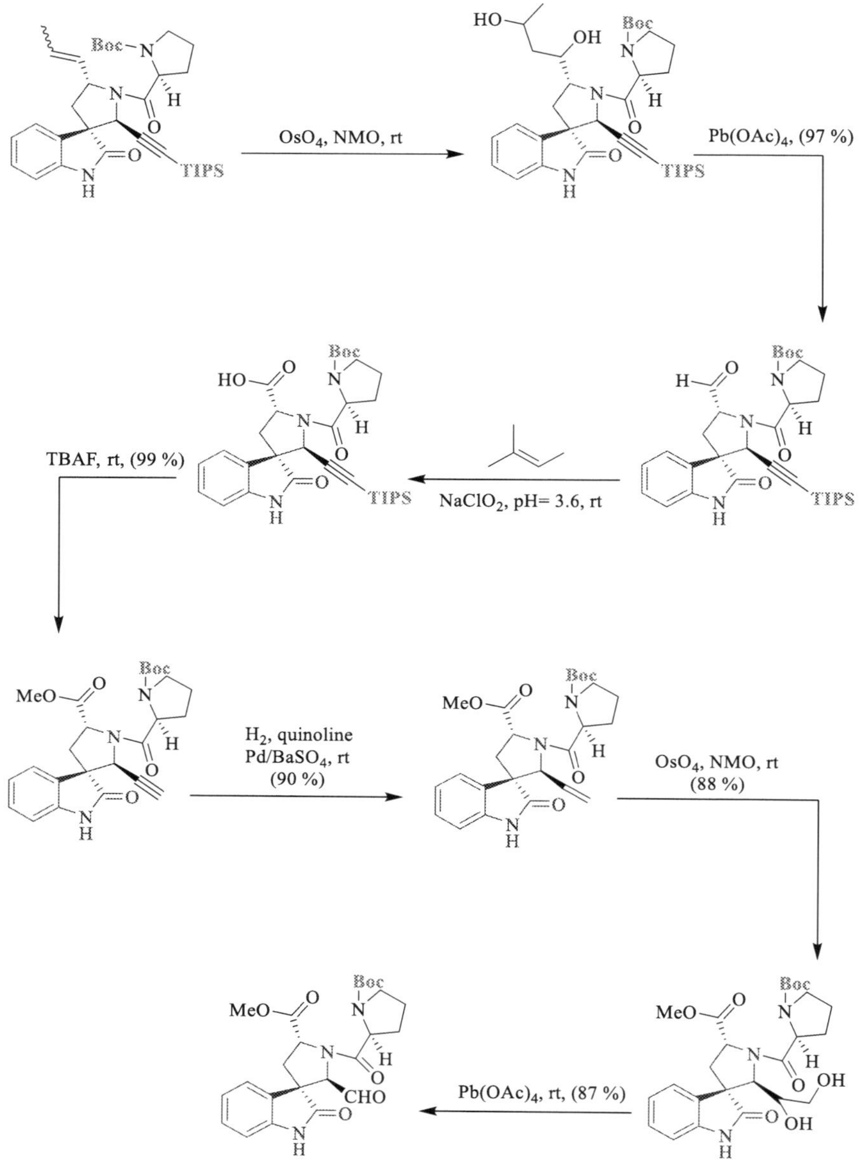

196
12. Spirotryprostatin B

SCHEME 12.7 (Continued).

Indole Alkaloids

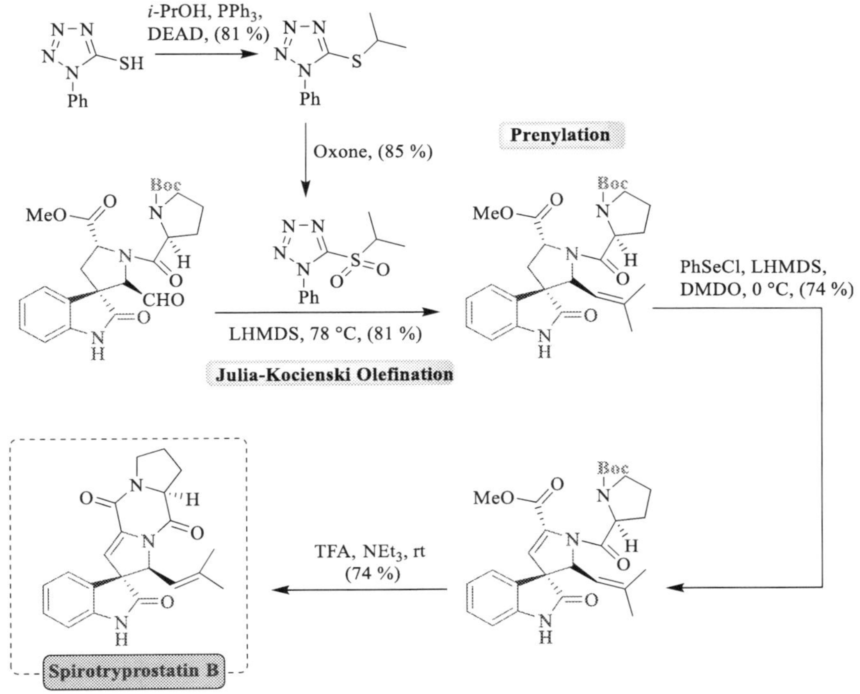

12.1 Total syntheses of (−)-spirotryprostatin B

197

SCHEME 12.7 (Continued).

Key features

- Synthesis of vinyl-substituted cyclopropylspirooxindole
- MgI$_2$-mediated ring expansion reaction of cyclopropylspirooxindole with an aldimine leading to rapid assembly of the spirotryprostatin core
- Condensation with *N*-Boc-L-proline
- Installation of prenyl side chain by Julia−Kocienski olefination of a key aldehyde precursor

Indole Alkaloids

12.1.8 Trost's total synthesis of spirotryprostatin B via diastereoselective prenylation (2007)

See Scheme 12.8.

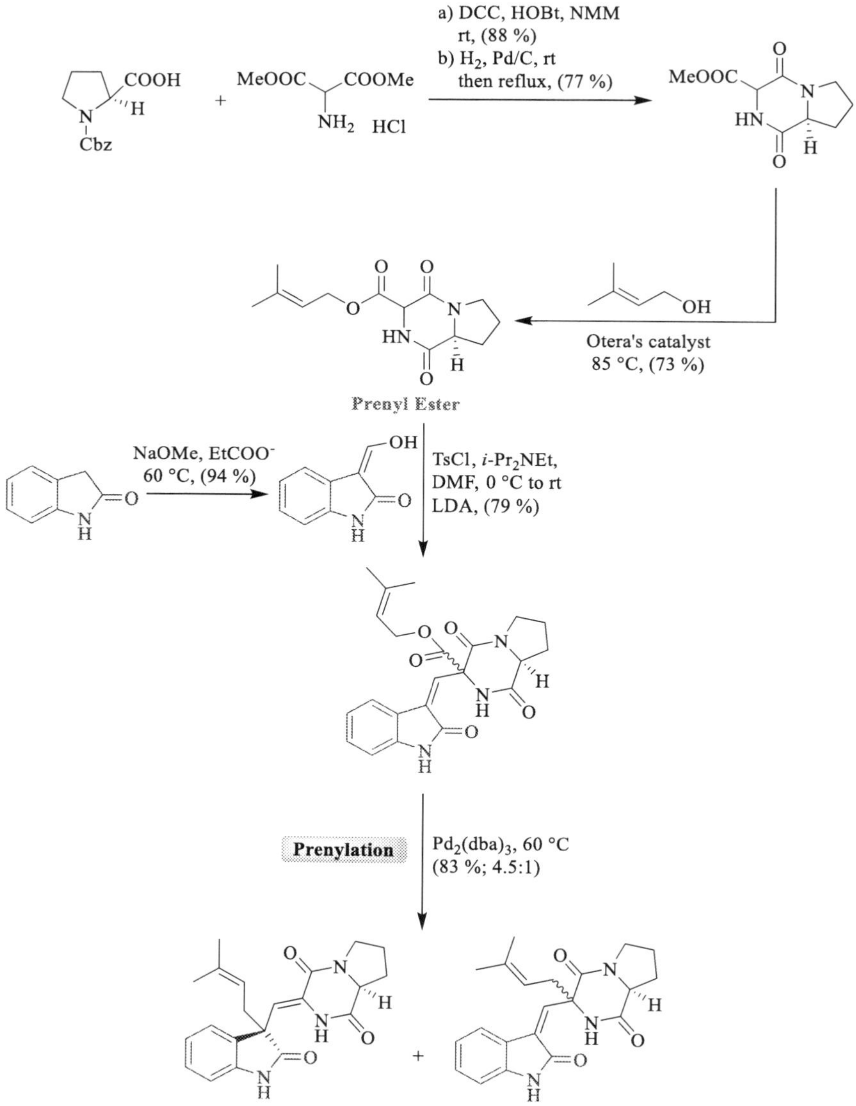

SCHEME 12.8 Trost's total synthesis of spirotryprostatin B via diastereoselective prenylation.[12]

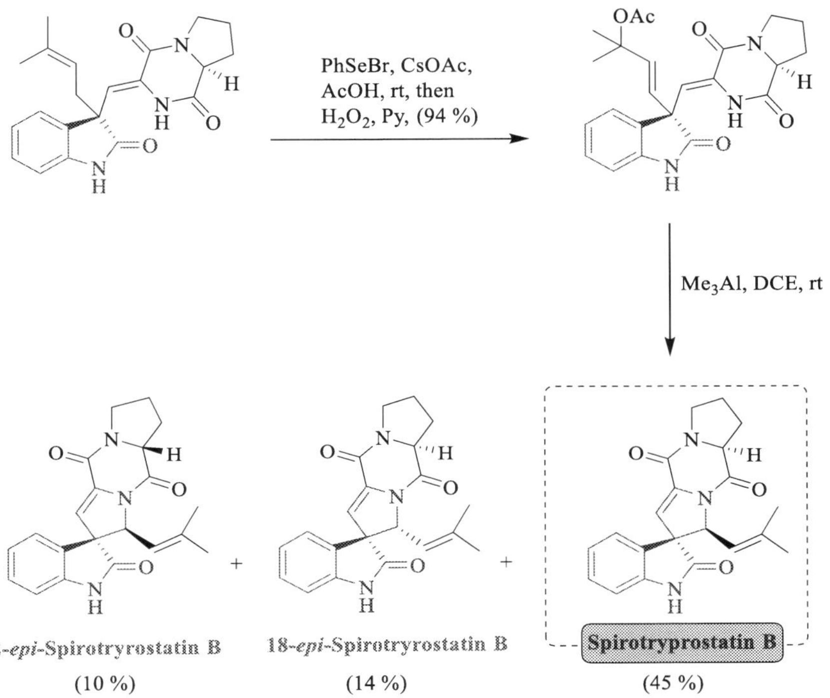

SCHEME 12.8 (Continued).

Key features

- Spirotryprostatin B was synthesized in eight steps in a 13% overall yield starting from oxindole.
- Palladium-catalyzed prenylation reaction is a key step to construct the quaternary C3 stereocenter.
- Decarboxylation—alkylation of substituted β-keto ester with transposition of the double bond to achieve the targeted product.

References

1. Cui, C.-B.; Kakeya, H.; Osada, H. Novel Mammalian Cell Cycle Inhibitors, Spirotryprostatins A and B, Produced by *Aspergillus fumigatus*, Which Inhibit Mammalian Cell Cycle at G2/M Phase. *Tetrahedron* **1996**, *52* (39), 12651–12666.
2. von Nussbaum, F.; Danishefsky, S. J. A Rapid Total Synthesis of Spirotryprostatin B: Proof of its Relative and Absolute Stereochemistry. *Angew. Chem. Int. Ed.* **2000**, *39* (12), 2175–2178.

200

3. Overman, L. E.; Rosen, M. D. Total Synthesis of (−)-Spirotryprostatin B and Three Stereoisomers. *Angew. Chem. Int. Ed.* **2000,** *39* (24), 4596−4599.
4. Overman, L. E.; Rosen, M. D. Terminating Catalytic Asymmetric Heck Cyclizations by Stereoselective Intramolecular Capture of η3-Allylpalladium Intermediates: Total Synthesis of (−)-Spirotryprostatin B and Three Stereoisomers. *Tetrahedron* **2010,** *66* (33), 6514−6525.
5. Wang, H.; Ganesan, A. A Biomimetic Total Synthesis of (−)-Spirotryprostatin B and Related Studies. *J. Org. Chem.* **2000,** *65* (15), 4685−4693.
6. Bagul, T. D.; Lakshmaiah, G.; Kawabata, T.; Fuji, K. Total Synthesis of Spirotryprostatin B via Asymmetric Nitroolefination. *Org. Lett.* **2002,** *4* (2), 249−251.
7. Sebahar, P. R.; Williams, R. M. The Asymmetric Total Synthesis of (+)- and (−)-Spirotryprostatin B. *J. Am. Chem. Soc.* **2000,** *122* (23), 5666−5667.
8. Sebahar, P. R.; Osada, H.; Usui, T.; Williams, R. M. Asymmetric, Stereocontrolled Total Synthesis of (+) and (−)-Spirotryprostatin B via a Diastereoselective Azomethine Ylide [1, 3]-Dipolar Cycloaddition Reaction. *Tetrahedron* **2002,** *58* (32), 6311−6322.
9. Miyake, F. Y.; Yakushijin, K.; Horne, D. A. Preparation and Synthetic Applications of 2-Halotryptophan Methyl Esters: Synthesis of Spirotryprostatin B. *Angew. Chem. Int. Ed.* **2004,** *43* (40), 5357−5360.
10. Meyers, C.; Carreira, E. M. Total Synthesis of (−)-Spirotryprostatin B. *Angew. Chem. Int. Ed.* **2003,** *42* (6), 694−696.
11. Marti, C.; Carreira, E. M. Total Synthesis of (−)-Spirotryprostatin B: Synthesis and Related Studies. *J. Am. Chem. Soc.* **2005,** *127* (32), 11505−11515.
12. Trost, B. M.; Stiles, D. T. Total Synthesis of Spirotryprostatin B via Diastereoselective Prenylation. *Org. Lett.* **2007,** *9* (15), 2763−2766.

CHAPTER
13

Strychnofoline

Strychnofoline (Fig. 13.1) is a strychnos alkaloid comprising unique spirooxindole architecture. It was first isolated by Angenot's research group from the leaves of *Strychnos usambarensis* in 1978.[1]

It possesses very promising antimitotic activity against cultures of mouse melanoma B16, Hepatom HW165, and Ehrlich tumor cells.

First total synthesis of this biologically important scaffold was described by Carreira in 2002.[2] They later modified and improved their racemic synthesis in 2006.[3] First enantioselective synthesis of (−)-strychnofoline was reported by Yu's research group.[4] All these synthetic approaches are elaborated in further sections.

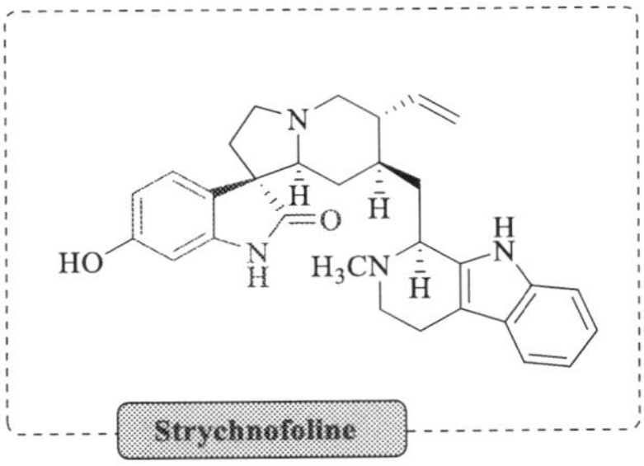

FIGURE 13.1 Structure of (−)-strychnofoline.

Indole Alkaloids
DOI: https://doi.org/10.1016/B978-0-323-91674-5.00011-0

© 2022 Elsevier Inc. All rights reserved.

202

13. Strychnofoline

13.1 Carreira's first total synthesis of (±)-strychnofoline (2002)

See Scheme 13.1.

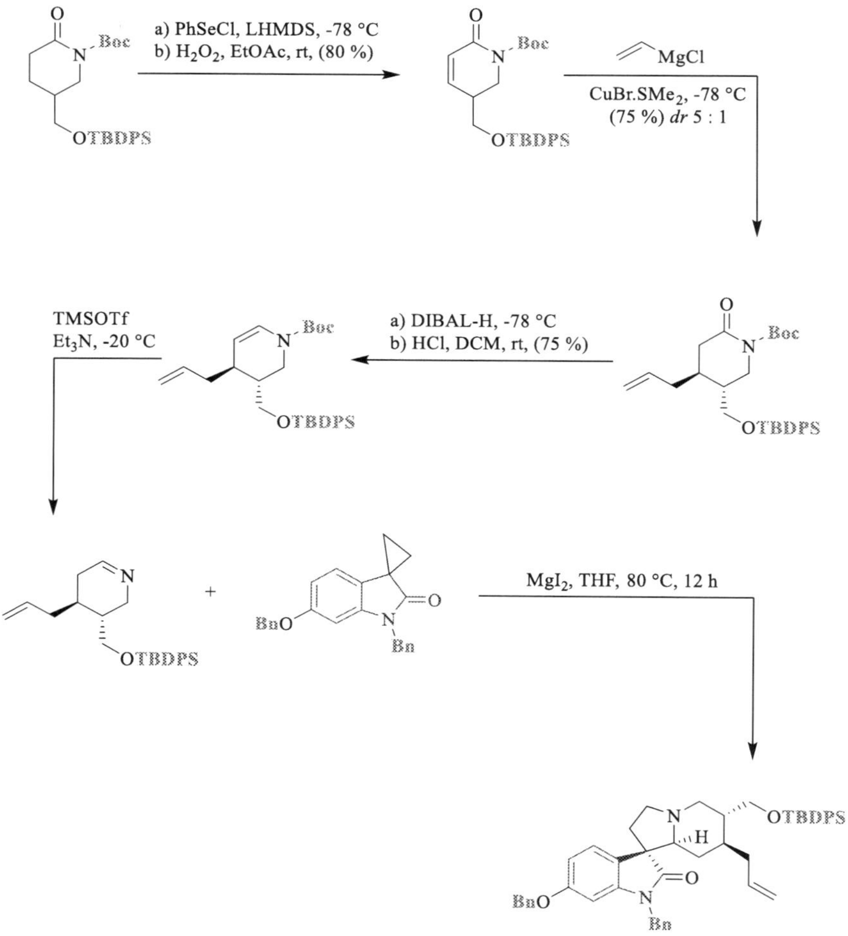

SCHEME 13.1   Carreira's first total synthesis of (±)-strychnofoline.[2]

Indole Alkaloids

13.1 Carreira's first total synthesis of (±)-strychnofoline (2002)

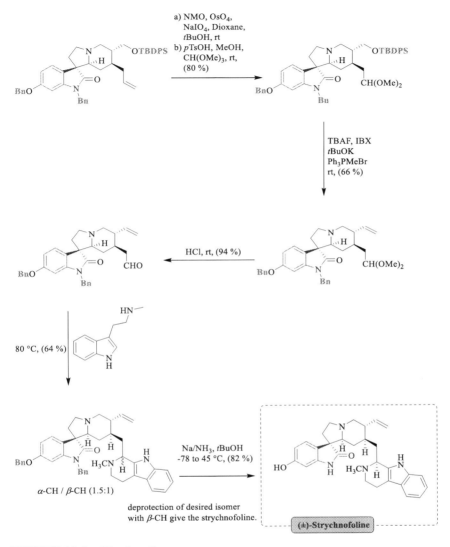

SCHEME 13.1 (Continued).

204
13. Strychnofoline

Key features

- First racemic total synthesis of antitumor alkaloid (±)-strychnofoline.
- Key synthetic step is the diastereoselective coupling of a cyclic imine with spiro[cyclopropan-1,3'-oxindole].
- Synthesis of spirocyclic cyclopropane.
- Annulation reaction; expansion of spiro cyclopropane to spiro cyclopentane; synthesis of four ringed spiro-structure.
- Synthetic routes for necessary side chain functionalities.
- Condensation of *N*-methyltryptamine to afford the final racemic (±)-strychnofoline.

13.2 Carreira's racemic total synthesis of (±)-strychnofoline; highly convergent selective annulation reaction (2006)

See Scheme 13.2.

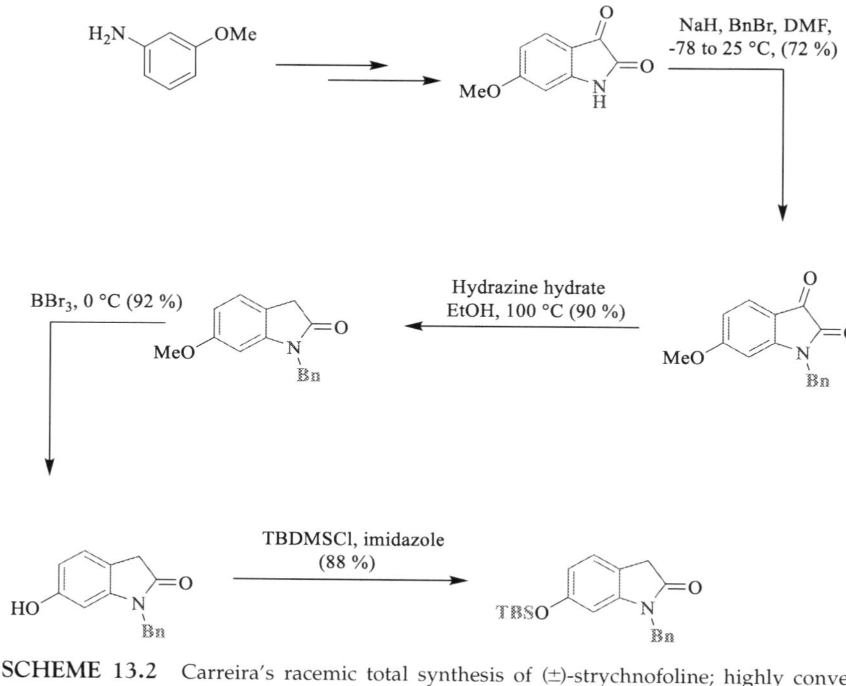

SCHEME 13.2 Carreira's racemic total synthesis of (±)-strychnofoline; highly convergent selective annulation reaction.[3]

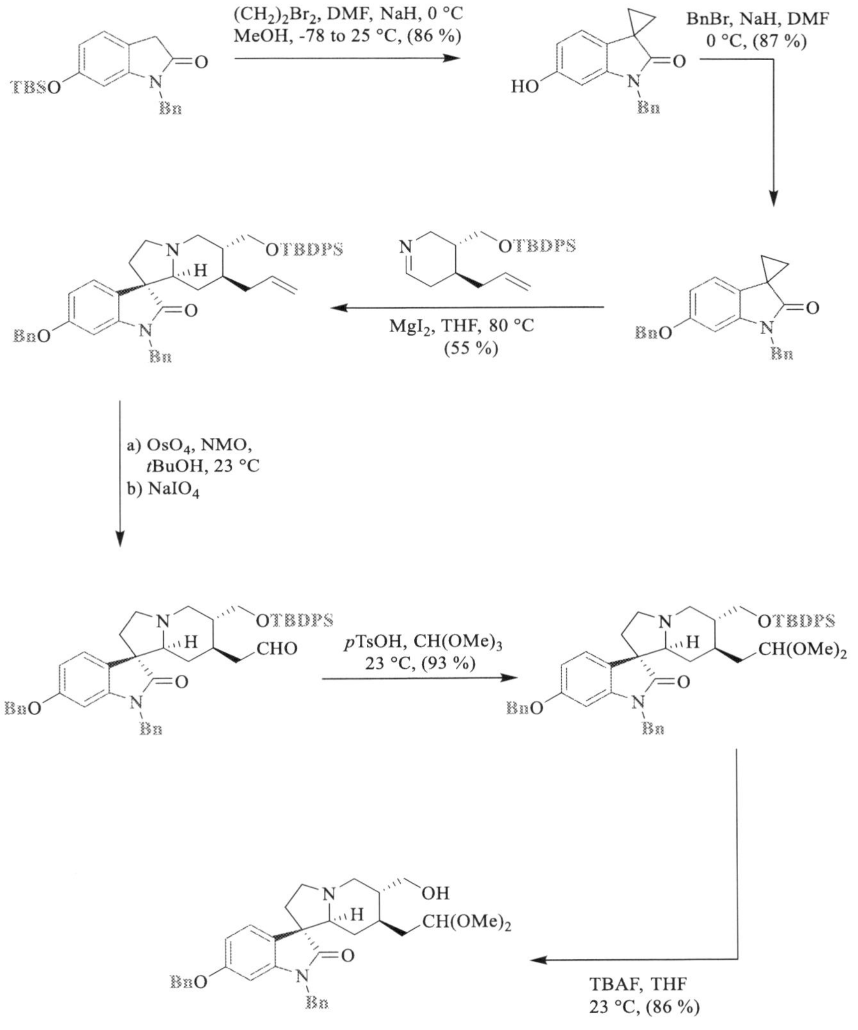

13.2 Carreira's racemic total synthesis of (±)-strychnofoline

205

SCHEME 13.2 (Continued).

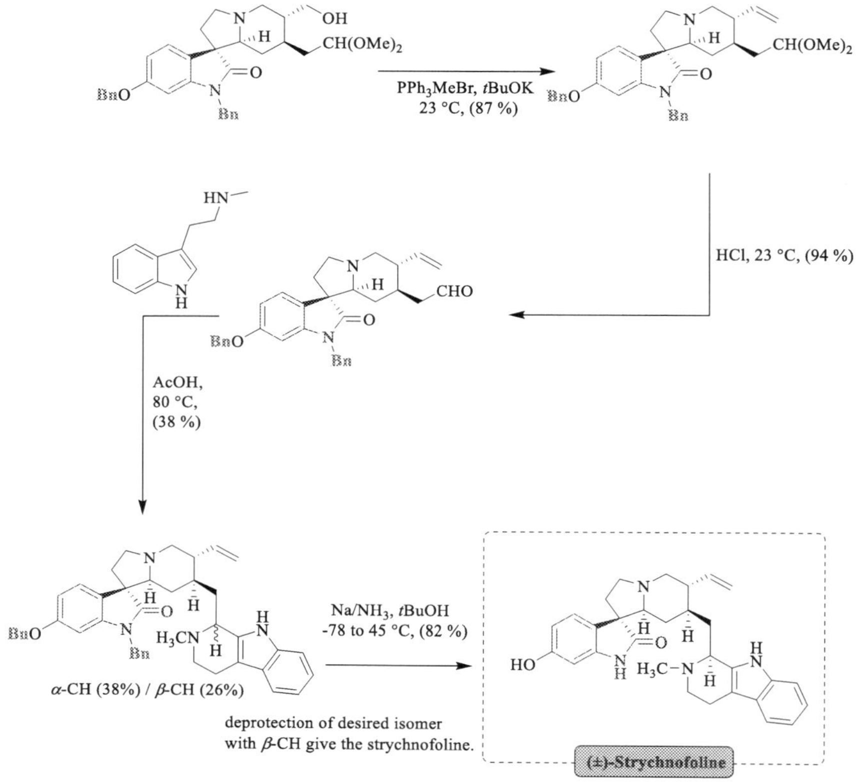

13. Strychnofoline

SCHEME 13.2 (Continued).

Key features

- Synthesis of spirocyclic cyclopropane starting from 6-methoxyisatin
- Annulation reaction; expansion of spiro cyclopropane to spiro cyclopentane; synthesis of four ringed spiro-structure
- Synthetic routes for necessary side chain functionalities
- Condensation of *N*-methyltryptamine to afford the final racemic (±)-strychnofoline

13.3 Yu's enantioselective synthesis of (−)-strychnofoline (2018)

See Scheme 13.3.

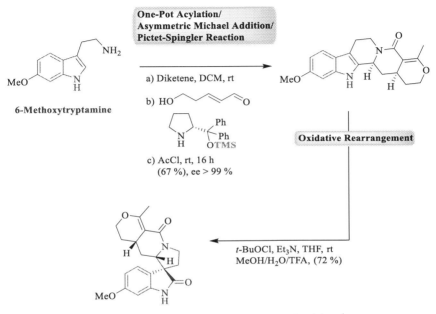

SCHEME 13.3 Yu's enantioselective synthesis of (−)-strychnofoline.[4]

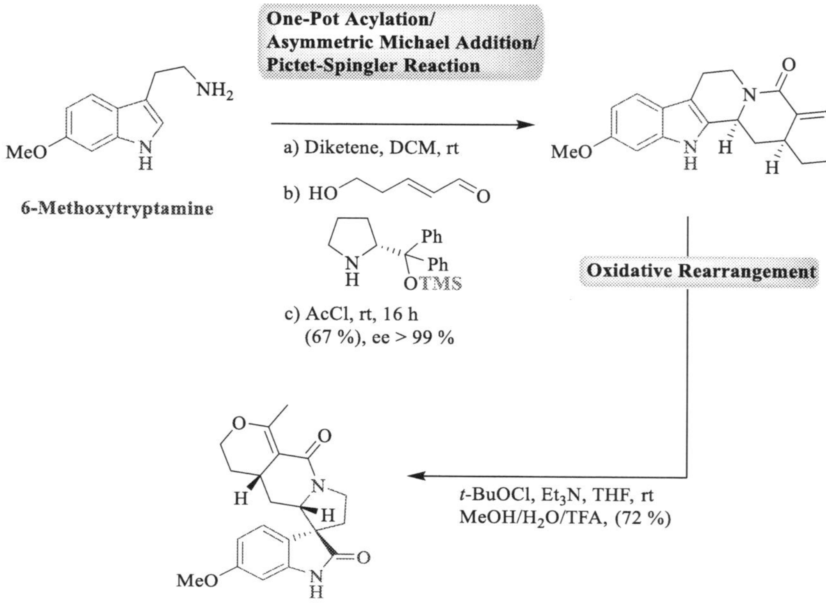

208   13. Strychnofoline

SCHEME 13.3 (Continued).

Key features

- Enantioselective (−)-strychnofoline was synthesized in nine steps starting from commercially available 6-methoxytryptamine.
- Development of one-pot, catalytic asymmetric synthetic protocol for quinolizidine skeleton and its subsequent rearrangement to spirooxindole in only two steps.
- One-pot acylation/asymmetric Michael addition/Pictet−Spengler reaction on 6-methoxytryptamine to afford five ring intermediate.
- Oxidative rearrangement to spiro-motif.
- Selective reduction of amide.
- One-pot Shapiro tosylhydrazone decomposition and deacylation.
- Oxidation and subsequent condensation with *N*-methyltryptamine.

References

1. Angenot, L. Nouveaux Alcaloïdes Oxindoliques du Strychnos Usambarensis GILG. *Plantes medicinales et Phytotherapie* **1978,** *12* (2), 123−129.
2. Lerchner, A.; Carreira, E. M. First Total Synthesis of (±)-Strychnofoline via a Highly Selective Ring-Expansion Reaction. *J. Am. Chem. Soc.* **2002,** *124* (50), 14826−14827.
3. Lerchner, A.; Carreira, E. M. Synthesis of (±)-Strychnofoline via a Highly Convergent Selective Annulation Reaction. *Chem. Eur. J.* **2006,** *12* (32), 8208−8219.
4. Yu, Q.; Guo, P.; Jian, J.; Chen, Y.; Xu, J. Nine-Step Total Synthesis of (−)-Strychnofoline. *Chem. Commun.* **2018,** *54* (9), 1125−1128.

List of abbreviations

AIBN	azoisobutyronitrile
Acac	acetylacetonate
Bn	benzyl
Boc	*t*-Butyloxycarbonyl
BOP	bis-(20-oxo-3-oxazolidinyl)phosphoryl
BRSM	based on recovered starting material
Bz	benzoyl
CAN	ceric ammonium nitrate
CSA	camphorsulfonic acid
DABCO	1,4-Diazabicyclo[2.2.2]-octane
dba	(*E,E*)-Dibenzylideneacetone
DBU	1,8-Diazabicyclo[5.4.0]undec-7-ene
DCC	dicyclohexylcarbodiimide
DCE	dichloroethane
DEAD	diethyl azodicarboxylate
DIBAL-H	diisobutylaluminum hydride
DIC	di-isopropylcarbodiimide
DIEA	*N,N*-Diisopropylethylamine
DMAP	4-Dimethylaminopyridine
DMC	dimethyl carbonate
DMDO	dimethyldioxirane
DME	1,2-Dimethoxyethane
DMP	Dess-Martin Periodinane
DMPU	1,3-Dimethyl-2-oxohexahydropyrimidine
DNBA	2-Methyl-6-nitrobenzoic acid anhydride
DPPA	diphenylphosphoryl chloride
DPPF	1,1'-Bis(diphenylphosphino)ferrocene
DTBMP	2,6-Di-tert-butyl-4-methylpyridine

Fmoc	fluorenylmethyloxycarbonyl chloride
HMDS	1,1,1,2,3,3,3-Hexamethyldisilazane
HMPA	hexamethylphosphoramide
IBX	2-Iodoxybenzoic acid
KHMDS	potassium hexamethyldisilazide
LiDBB	lithium 4,4'-di-*tert*-butylbiphenylide
LHMDS	lithium hexamethyldisilazide
MAD	Methylaluminum bis(2,6-di-*tert*-butyl-4-methylphenoxide
m-CPBA	*meta*-Chloroperbenzoic acid
MOM	methhoxymethyl ether
Ms	methanesulfonyl
MTPI	methyltriphenoxyphosphonium iodide
NaHMDS	sodium hexamethyldisilazide
NBS	*N*-Bromosuccinamide
NCS	*N*-Chlorosuccinamide
NMO	*N*-Methylmorpholine-*N*-oxide
o-tol	*o*-Tolyl
PMB	*p*-Methoxybenzyl ether
PMP	1,2,2,6,6-Pentamethylpiperidine
PPTS	pyridinium *p*-toluenesulfonate
Py	pyridine
Red-Al	sodium bis(2-methoxyethoxy)aluminium hydride
rt	room temperature
SEM	2-(Trimethylsilyl)ethoxymethyl chloride
TBAF	tetra-*n*-butylammonium fluoride
TBAT	tetrabutylammonium difluorotriphenylsilicate
TBDMS	*t*-Butyldimethylsilyl
TBDPS	*t*-Butyldiphenylsilyl
TBS	*t*-Butylsilyl
TBTH	tributyltin hydride
TCC	trans-2-(α-cumyl)cyclohexanol
TEMPO	2,2,6,6-Tetramethylpiperidine-*N*-oxyl
TES	triethylsilyl

Tf	trifluoromethansulfonyl
TFA	trifluroacetic acid
TFAA	trifluoroacetic anhydride
TIPS	triisopropysilyl ether
TMS	trimethylsilyl
TMEDA	N,N,N',N'-tetramethylethylenediamine
Troc	2,2,2-Trichloroethoxycarbonyl
WSC	N-Ethyl-N'-(3-dimethylaminopropyl)carbodiimide hydrochloride

Index

Note: Page numbers followed by "*f*" refer to figures.

A

(−)-affinisine, 5
  Fonseca's first stereospecific total
    synthesis of, 6−7, 7*f*
  Fonseca's improved total synthesis of,
    8−10, 10*f*
  structure of, 5*f*
α,β-unsaturated lactams, 162
Amide coupling, 162
Aminal urethane, 105
Asymmetric allylic alkylation, 71
Asymmetric nitroolefination, 62
Asymmetric Pictet−Spengler/Dieckmann-
    condensation, 7, 10
Asymmetric synthesis, 136−145, 141*f*, 145*f*
Asymmetric total synthesis, 176−177, 177*f*
(+)-Austamide, 13
  biosynthesis of, 13, 14*f*
  Corey's total synthesis of (+)-austamide,
    18−20
  Hutchison's total synthesis of
    (±)-austamide, 14−18, 15*f*
  structure of, 13*f*
Aza-Michael initiated ring closure (aza-
    MIRC), 80
Azomethine ylides, 181

B

Ban's intermediate, 157, 167, 171
Barton-modified Hunsdiecker protocol, 193
Beckmann rearrangement, 180
Borschberg's racemic total synthesis of
    (+)-elacomine and (−)-isoelacomine,
    88−89, 88*f*, 89*f*
Bosch chiral lactam, 157
Brevianamide A, 1*f*, 23
  biosynthesis of, 24*f*
  Lawrence's total synthesis of, 28−31, 30*f*
  structures of, 23*f*
Brevianamide B, 23
  biosynthesis of, 24*f*
  Lawrence's total synthesis of, 28−31, 30*f*

Scheerer's formal synthesis of, 31−33, 32*f*
  structures of, 23*f*
  Williams' asymmetric total synthesis of,
    25−28
Brevianamide C, 1*f*
Brevianamide F, 24−25
Butenylation, 155

C

Carbamoylation, 40
Carboline, 64, 84
Carbomagnesation, 162
Carreira, 201−206, 202*f*, 204*f*
Carreira's diastereoselective [3 + 2]
    annulation, 156−158
Carreira's racemic total synthesis of
    (±)-horsfiline, 63, 63*f*
Castillo's formal synthesis of
    (±)-coerulescine, 84, 84*f*
Chang's racemic total synthesis of
    (±)-coerulescine, 69−70, 69*f*
Chemoselective reduction, 167
Chiral auxiliary, 48, 157
Chiral Bronsted acid, 181
Citrinadin A, 1*f*, 35
  biosynthetic pathway, 35, 36*f*
  Martin's enantioselective total syntheses
    of (−)-Citrinadin A and
    (+)-Citrinadin B, 47−51, 48*f*
  Martin's first enantioselective total
    synthesis of Citrinadin A, 43−47, 44*f*
  Martin's synthesis of Citrinadin A
    spirooxindole core, 35−38, 37*f*
  structure of, 36*f*
  Wood's enantioselective synthesis of
    (+)-Citranidin A core, 41−43, 41*f*
Citrinadin B, 1*f*, 35
  biosynthetic pathway, 35, 36*f*
  Sorensen's synthesis of Citrinadin B core
    architecture, 39−40, 39*f*
  structure of, 36*f*
  synthesis of (+)-Citrinadin B, 49−51, 50*f*

213

214 Index

Citrinadin B (*Continued*)
  Wood's enantioselective total synthesis of (+)-Citrinadin B, 51–55, 52*f*
Claisen condensation, 158
Claisen rearrangement, 76
Coerulescine, 1*f*, 3, 57
  Castillo's formal synthesis of (±)-coerulescine, 84, 84*f*
  Chang's racemic total synthesis of (±)-coerulescine, 69–70, 69*f*
  Comesse's racemic total synthesis of (±)-coerulescine, 80, 80*f*
  Hayashi's enantioselective total synthesis of (−)-coerulescine and (−)-horsfiline, 83–84, 83*f*
  Kim's total synthesis of (−)-coerulescine, 79, 79*f*
  Kulkarni's total synthesis of (±)-coerulescine and (±)-horsfiline, 75–76, 75*f*, 76*fa*
  Masanori's total synthesis of (±)-coerulescine, 64, 64*f*
  Murphy's racemic total synthesis of (±)-coerulescine and (±)-horsfiline, 67–68, 67*f*, 68*fa*
  Selvakumar's racemic total synthesis of (±)-coerulescine and (±)-horsfiline, 66, 66*f*
  structures of, 57*f*
  White's racemic total synthesis of (±)-coerulesine and (±)-horsfiline, 77–78, 77*f*, 78*fa*
Comesse's racemic total synthesis of (±)-coerulescine, 80, 80*f*
Convergent selective annulation reaction, 204–206, 204*f*
Corey–Chaykovsky epoxidation, 53
Corey's total synthesis of (+)-austamide, 18–20
Corynoxeine, 155
Cross-dehydrogenative coupling (CDC), 170–171
Cross-metathesis, 167
Curtius rearrangement, 104*f*, 105, 146
Cyclization, 66, 68, 71, 76
Cyclopropanation, 110

D

Danishefsky's reverse prenylation, 31
Davis' oxaziridine, 51
Deformylation, 79
Dehydrobrevianamide E, 31

Dehydrospiropyrrolidine, 183
Dieckmann, 59
  condensation, 7, 153
Dihydrocorynantheine, 159
Dihydrosecologanin aglycone, 159–160, 160*f*
Dihydrospirotryprostatin B, 189, 195
Diketopiperazine, 141, 145, 173, 175, 180, 185, 187, 193, 195
DMDO-mediated oxidative rearrangement, 38

E

(*E*)-dienoate, 187
8-phenylmenthol, 38
Elacomine, 1*f*, 3–4, 87
  Borschberg's racemic total synthesis of (+)-elacomine and (−)-isoelacomine, 88–89, 88*f*, 89*f*
  Horne's racemic total synthesis of (±)-elacomine and (±)-isoelacomine, 90–91, 90*f*, 91*fa*
  Njardarson's asymmetric total synthesis of (+)-elacomine, 94, 94*f*
  structure of, 87*f*
  Takemoto's formal synthesis of elacomine and isoelacomine, 95–97, 95*f*, 96*fa*
  White's racemic total synthesis of (±)-elacomine, 92–93, 92*f*, 93*fa*
Ellman imine, 94
Ender's epoxidation, 55
Enone, 43, 51, 53, 55
Eschweiler–Clarke, 60
η^3-allylpalladium, 187

F

5-exo cyclization, 60
Fischer-indole synthesis, 38
Fonseca's first stereospecific total synthesis of (−)-affinisine, 6–7, 7*f*
Fonseca's improved total synthesis of (−)-affinisine, 8–10, 10*f*
Formal synthesis, 156*f*, 161–171, 161*f*, 163*f*, 165*f*, 168*f*, 170*f*
  of gelsemine, 122–134
Fuji's asymmetric total synthesis of (−)-horsfiline, 61–63, 61*f*, 62*fa*
Fukuyama's enantioselective total synthesis of (+)-gelsemine, 105–110, 109*f*
Fukuyama's first racemic total synthesis of (±)-gelsemine, 100–105, 104*f*

Index

G

Gassman oxindole synthesis, 146
Gelsemine, 99
 formal syntheses of, 122–134
  Johnson's spiroannelation of gelsemine, 126–127, 127f
  Johnson's synthesis of key tetracyclic gelsemine intermediate, 122–126, 125f
  Zhou's second attempt for spirocyclopentaneoxindole intermediate through intramolecular Michael cyclization, 131–134, 133f
  Zhou's synthesis of spirocyclopentaneoxindole intermediate through intramolecular Michael cyclization, 128–131, 130f
 structure of, 99f
 total syntheses of, 100–122
  Fukuyama's enantioselective total synthesis of (+)-gelsemine, 105–110, 109f
  Fukuyama's first racemic total synthesis of (±)-gelsemine, 100–105, 104f
  Overman's total synthesis of (±)-gelsemine, 110–114, 113f
  Qin's total synthesis of (+)-gelsemine, 114–118, 117f
  Qiu's asymmetric total synthesis of (+)-gelsemine, 118–122, 121f
Gelsemine, 1f
Gelsemium alkaloids, 1
Gramine, 141

H

Hayashi's enantioselective total synthesis of (−)-coerulescine and (−)-horsfiline, 83–84, 83f
Heck cyanation, 73
Heck cyclization, 94
Heck reaction, 114, 173, 180
Heptacycle, 141
Horne's racemic total synthesis of (±)-elacomine and (±)-isoelacomine, 90–91, 90f, 91fa
Horner–Wordsworth–Emmons, 104f, 109f, 110, 153, 164
Horsfiline, 1f, 3–4, 57
 Carreira's racemic total synthesis of (±)-horsfiline, 63, 63f

Fuji's asymmetric total synthesis of (−)-horsfiline, 61–63, 61f, 62fa
Hayashi's enantioselective total synthesis of (−)-coerulescine and (−)-horsfiline, 83–84, 83f
Jones' total synthesis of horsfiline via radical cyclization, 58–60, 58f, 60fa
Kulkarni's total synthesis of (±)-coerulescine and (±)-horsfiline, 75–76, 75f, 76fa
Maison's total synthesis of (±)-horsfiline, 74
Murphy's racemic total synthesis of (±)-coerulescine and (±)-horsfiline, 67–68, 67f, 68fa
Neuville and Zhu's total synthesis of horsfiline, 72–73, 72f, 73fa
Palmisano's asymmetric total synthesis of (−)-horsfiline, 65, 65f
Park's enantioselective total synthesis of (−)-horsfiline, 81–82, 81f, 82fa
Selvakumar's racemic total synthesis of (±)-coerulescine and (±)-horsfiline, 66, 66f
structures of, 57f
Trost's asymmetric total synthesis of (−)-horsfiline, 70–71, 70f, 71fa
White's racemic total synthesis of (±)-coerulescine and (±)-horsfiline, 77–78, 77f, 78fa
Hutchison's total synthesis of (±)-austamide, 14–18, 15f
Hydrogenolysis, 18
 deprotection, 89

I

Intermolecular [3 + 2] nitrone cycloaddition, 42
Intramolecular cyclization, 28, 185
Intramolecular iminium, 91
Intramolecular lactamization, 82
Intramolecular oxymercuration, 104f, 105, 109f
Intramolecular $S_N 2$ displacement, 122
Ion spirocyclization, 91
Isoelacomine, 1f, 3, 87
 Borschberg's racemic total synthesis of (+)-elacomine and (−)-isoelacomine, 88–89, 88f, 89f
 Horne's racemic total synthesis of (±)-elacomine and (±)-isoelacomine, 90–91, 90f, 91fa

216

Index

Isoelacomine (*Continued*)
  structure of, 87*f*
  Takemoto's formal synthesis of
    elacomine and isoelacomine, 95–97,
    95*f*, 96*fa*
Isorhynchophyline, 1*f*

J

Johnson's spiroannelation of gelsemine,
  126–127, 127*f*
Johnson's synthesis of key tetracyclic
  gelsemine intermediate, 122–126,
  125*f*
Jone's oxidation, 69, 76, 109*f*, 110
Jones' total synthesis of horsfiline via
  radical cyclization, 58–60, 58*f*, 60*fa*
Julia–Kocienski olefination, 197

K

Kim's total synthesis of (−)-coerulescine,
  79, 79*f*
Knoevengel condensation, 104*f*, 109*f*
Kulkarni's total synthesis of
  (±)-coerulescine and (±)-horsfiline,
  75–76, 75*f*, 76*fa*

L

Lactam, 40, 51
Lactamization, 157
Lawrence's total synthesis of
  Brevianamides A and B, 28–31, 30*f*
L-proline, 180

M

Maison's total synthesis of (±)-horsfiline, 74
Mannich spirocyclization, 155
Mannich–Michael addition, 163*f*
Marcfortine A, 141
Martin's enantioselective total syntheses of
  (−)-Citrinadin A and (+)-Citrinadin
  B, 47–51, 48*f*
Martin's first enantioselective total
  synthesis of Citrinadin A, 43–47, 44*f*
Masanori's total synthesis of
  (±)-coerulescine, 64, 64*f*
Methoxyvinylation, 169
Michael addition, 59, 83, 104*f*, 105, 109*f*,
  110, 117*f*, 118, 131, 134, 157, 167
Michael–Krapcho sequence, 170–171
Mixed Claisen acylation, 40
Monomethylation, 55
Mukaiyama aldol reaction, 180

Murphy's racemic total synthesis of
  (±)-coerulescine and (±)-horsfiline,
  67–68, 67*f*, 68*fa*

N

N-acyl-N,O-hemiacetal, 18
Neuville and Zhu's total synthesis of
  horsfiline, 72–73, 72*f*, 73*fa*
Njardarson's asymmetric total synthesis of
  (+)-elacomine, 94, 94*f*

O

One pot, 162, 170–171
[1,3]-dipolar cycloaddition, 66, 177
Overman's total synthesis of (±)-gelsemine,
  110–114, 113*f*
Oxidation, 18
Oxidative rearrangement, 173, 175
Oxindole alkaloids, 1
Ozonolysis, 122

P

Palmisano's asymmetric total synthesis of
  (−)-horsfiline, 65, 65*f*
Paraherquamide A/Paraherquamide B, 1*f*
  conversion of Marcfortine A to, 146–149,
  147*f*
  total syntheses, 136–145
    McWhorter's formal synthesis,
    145–146, 146*f*
    Williams and Cushing convergent,
    stereocontrolled, asymmetric total
    synthesis, 142–145, 145*f*
    Williams' asymmetric stereocontrolled
    first total synthesis, 136–141, 141*f*
Park's enantioselective total synthesis of
  (−)-horsfiline, 81–82, 81*f*, 82*fa*
Pfitzner–Moffatt oxidation, 104*f*, 105
Pictet–Spengler condensation, 89, 189
Pictet–Spengler reaction, 175
Pinnick oxidation, 157–158, 167
Prenylation, 198–199, 198*f*
PTC allylation, 82

Q

Qin's total synthesis of (+)-gelsemine,
  114–118, 117*f*
Qiu's asymmetric total synthesis of
  (+)-gelsemine, 118–122, 121*f*

R

Racemic, 201, 204–206, 204*f*

Index

217

Radical cyclization, 58–60, 58f
Rearrangement, 18
Reductive amination, 71
Retro-Mannich fragmentation, 78, 87, 93
Reverse prenylation, 24–25
Rhynchophylline/isorhynchophylline
  formal syntheses, 161–171
    Amat's enantioselective formal
      synthesis, 168–169, 168f
    Itoh's enantioselective formal
      synthesis, 162–164, 163f
    Martin's racemic formal synthesis,
      161–162, 161f
    Wang's enantioselective formal
      synthesis, 165–167, 165f
    Xia's racemic formal synthesis,
      170–171, 170f
  semisyntheses of, 159–160
    Brown's semisynthesis, 159–160, 160f
    Finch's semisynthesis, 159, 159f
  total syntheses, 151–158
    Ban's first total synthesis, 152–153,
      152f
    Hiemstra's total synthesis, 154–155,
      154f
    Tong's total synthesis, 156–158, 156f,
      158f
Ring-closing metathesis (RCM), 161–162,
  161f

S

Scheerer's formal synthesis of
  Brevianamide B, 31–33, 32f
Selvakumar's racemic total synthesis of
  (±)-coerulescine and (±)-horsfiline,
  66, 66f
Semisynthesis, 159–160, 159f, 160f
Somei–Kametani coupling, 141, 145
Sonogashira coupling, 47, 51
Spiro cyclopropanation, 63
Spirooxindoles, 1, 1f, 3–4
  core, 35–43, 37f
Spirotryprostatin A, 1f, 3–4
  total synthesis, 174–181
    Danishefsky's total synthesis, 174–175
    Fukuyama's stereoselective total
      synthesis, 178–180
    Gong's synthesis, 181
    Williams' asymmetric total synthesis,
      176–177
Spirotryprostatin B, 1f, 3–4
  total syntheses of, 184–199

Biomimetic total synthesis, 187–189,
  188f
Carreira's total synthesis, 195–197,
  195f
Danishefsky's total synthesis,
  184–185
Fuji's total synthesis, 190–191, 190f
Horne's total synthesis, 194–195, 194f
Rosen's total synthesis, 186–187, 186f
Trost's total synthesis, 198–199, 198f
Williams' asymmetric stereocontrolled
  total synthesis, 192–193, 192f
Strychnofoline
  Carreira's first total synthesis of,
    201–204, 202f
  Carreira's racemic total synthesis of,
    204–206, 204f
  structure of, 201f
  Yu's enantioselective synthesis of,
    207–208, 207f

T

Takemoto's formal synthesis of elacomine
  and isoelacomine, 95–97, 95f, 96fa
[3 + 2] Annulation, 65, 94
[3 + 2] cycloaddition, 181
Total synthesis, 141f, 142–145, 145f,
  183–199, 184f, 186f, 188f, 190f, 192f,
  194f, 195f, 198f, 201–206, 202f, 204f.
  See also specific type
  of (±)-elacomine and (±)-isoelacomine,
    88–97
  of coerulescine, 58–84
  of gelsemine, 100–122
  of horsfiline, 58–84
Transfer hydrogenation, 60
Trost's asymmetric total synthesis of
  (−)-horsfiline, 70–71, 70f, 71fa
Tryptamine, 155
Tsuji allylation, 79
12-*epi*-Spirotryprostatin B, 193
[2 + 2] photocycloaddition, 78, 93

W

White's racemic total synthesis of
  (±)-coerulescine and (±)-horsfiline,
  77–78, 77f, 78fa
White's racemic total synthesis of
  (±)-elacomine, 92–93, 92f, 93fa
Williams' asymmetric total synthesis of
  Brevianamide B, 25–28, 27f
Wittig reaction, 121f, 122

Wood's enantioselective total synthesis of (+)-Citrinadin B, 51−55, 52*f*

Y
Yamada modification, 146

Z
Zhou's second attempt for spirocyclopentaneoxindole intermediate through intramolecular Michael cyclization, 131−134, 133*f*

Zhou's synthesis of spirocyclopentaneoxindole intermediate through intramolecular Michael cyclization, 128−131, 130*f*

Printed in the United States
by Baker & Taylor Publisher Services